IS GLOBAL
WARMING
A THREAT?

Other books in the At Issue series:

IS GLOBAL WARMING A THREAT?

Mary E. Williams, *Book Editor*

Daniel Leone, *President*
Bonnie Szumski, *Publisher*
Scott Barbour, *Managing Editor*

GREENHAVEN
PRESS ®

THOMSON
™
GALE

San Diego • Detroit • New York • San Francisco • Cleveland
New Haven, Conn. • Waterville, Maine • London • Munich

For more information, contact
Greenhaven Press
27500 Drake Rd.
Farmington Hills, MI 48331-3535
Or you can visit our Internet site at http://www.gale.com

LIBRARY OF CONGRESS CATALOGING-IN-PUBLICATION DATA

Is global warming a threat? / Mary E. Williams, book editor.
    p. cm. — (At issue)
Includes bibliographical references and index.
ISBN 0-7377-1333-X (pbk. : alk. paper) — ISBN 0-7377-1332-1 (lib. : alk. paper)
    1. Global warming—Environmental aspects. I. Williams, Mary E., 1960– .
II. At issue (San Diego, Calif.)
QC981.8.G56 I78 2003
363.738'74—dc21   2002027154

Printed in the United States of America

Contents

Introduction

In 1986, a panel of 150 scientists from eleven countries issued a report warning that human activities such as automobile use, the production of energy from burning fossil fuels, and deforestation could cause global temperatures to rise by intensifying the earth's greenhouse effect.

An essential component of the earth's climate, the greenhouse effect is the warming process that results from the atmospheric presence of heat-trapping gases such as water vapor, carbon dioxide, and nitrous oxide. Much of the solar energy that reaches the planet is absorbed by oceans and land masses, which in turn radiate the energy back into space. However, small concentrations of water vapor and other "greenhouse gases" convert some of this energy to heat and either retain it or reflect it back to the earth's surface. This "trapped" energy creates a blanket of warm air around the earth that moderates global temperatures and climate patterns. Without greenhouse gases, the earth would exist in a perpetual ice age.

The scientists who maintained in the 1980s that human activities could amplify the greenhouse effect were elaborating on a nineteenth-century theory proposed by Swedish chemist Svante Arrhenius. In 1896, Arrhenius hypothesized that the carbon dioxide produced from the burning of coal and other fossil fuels would cause global temperatures to rise by trapping excess heat in the earth's atmosphere. But the global warming theory did not capture the world's attention until 1988, when James Hansen, an atmospheric scientist and director of the Goddard Institute for Space Studies at NASA, testified before a U.S. Senate committee that the "evidence is strong" that human-made pollutants were raising world temperatures. If temperatures continued to rise, he warned, the earth would face catastrophic climate changes that would adversely affect the environment and human health.

Initially, most climate researchers were skeptical about Hansen's warning. It was true that carbon dioxide levels had increased by about 30 percent since the mid-1700s, when the Industrial Revolution began; it was also true that the average world temperature had risen by one degree Fahrenheit (F) during the twentieth century—the largest increase of any century during the past millenium. Yet the earth's climate had been prone to fluctuations over the past several hundred years, many climatologists maintained. The one degree temperature change could be attributed to the natural variability of the planet's weather.

As scientists conducted more climatological research, however, data supporting Arrhenius's global warming theory mounted. In 1995, the Intergovernmental Panel on Climate Change (IPCC), a United Nations subcommittee comprising more than two thousand scientists, asserted that human activities were partly responsible for rising global temperatures. In its ensuing assessment reports, the IPCC also predicted that carbon dioxide levels could double by the year 2100, causing temperatures to increase from 2 to 10.4 degrees F. Such a temperature change would likely bring a greater incidence of floods, droughts, heat waves, wildfires, hurricanes, and other forms of extreme weather—which in turn could cause an increase in storm-related deaths, infectious diseases, and economic crises, analysts warned. The weather of the 1990s—the hottest decade on record—seemed to bear these warnings out. As environmental journalist Ross Gelbspan points out, the year 1998 "began with a January ice storm that left four million people without power in Quebec and northern New England. For the first time, rainforests in Brazil and Mexico actually caught fire. The summer brought killer heat waves in the Middle East, India and Texas, where residents suffered through a record 29 consecutive triple-digit days." The following year was even worse, contends Gelbspan: "1999 saw a record-setting drought in the Mid-Atlantic states. . . . A heat wave in the Midwest and northeastern U.S. claimed 271 lives. Hurricane Floyd visited more than $1 billion in damages on North Carolina. A super-cyclone in eastern India killed 10,000 people. That winter, mudslides and rains in Venezuela claimed 15,000 lives. Unprecedented December windstorms swept northern Europe, causing more than $4 billion in damages."

Even before these extreme weather events occurred, a United Nations conference met in Kyoto, Japan, to discuss how to respond to the potential risks posed by global warming. In December 1997, UN negotiators approved an agreement requiring thirty-eight industrialized nations to reduce their greenhouse gas emissions by 6 to 8 percent below 1990 levels by the year 2012. Developing countries, which are poorer and less able to reduce greenhouse gases without straining their economies, were exempted from the emissions reduction requirement but were given voluntary standards as goals. Emmissaries signing this treaty—known as the Kyoto Protocol—must obtain approval from their own governments to render it a binding agreement. A Clinton administration official signed the Kyoto Protocol in 1998, but the U.S. Senate voted in 1999 to reject any climate change treaty that does not require poor nations to reduce their own greenhouse gases.

Although American environmentalists continue to advocate for the ratification of the Kyoto Protocol, critics argue that the treaty would hurt the economy and lower U.S. living standards. Cutting down greenhouse gas emissions would require the energy and industrial sectors to adopt expensive pollution-reducing technology—or to slow down their rate of production. The ensuing reduction in energy and goods would in turn raise gas, food, and housing costs, anti-Kyoto forecasters predict. Moreover, economists warn that adopting the protocol would place U.S. businesses at a competitive disadvantage if developing nations remained exempt from pollution-reducing requirements. Faced with the high cost of reducing emissions, some U.S. companies would relocate to developing countries, resulting in job losses for Americans.

Heeding these warnings from critics of the Kyoto Protocol, President George W. Bush officially withdrew U.S. support for the treaty in 2001. In a speech he delivered a few months after taking office, Bush also announced that he had formed a Cabinet-level working group to review the most up-to-date information on global warming and climate change. The working group contacted the highly respected National Academy of Sciences, which issued a report stating that the increase in carbon dioxide levels since the Industrial Revolution was largely due to human activity. However, noted Bush, "the Academy's report tells us that we do not know how much effect natural fluctuations in climate may have had on warming. We do not know how much our climate could, or will change in the future. We do not know how fast change will occur, or even how some of our actions could impact it."

Bush's speech reflects the skepticism that some scientists have about the global warming theory. Some climatologists maintain that the higher incidence of severe weather during the 1990s was not necessarily linked to higher greenhouse gas levels or higher global temperatures. Moreover, as several researchers point out, the IPCC's predictions about global warming are largely based on computer-generated climate simulations, which have proved to be unreliable. Global temperature readings taken on the ground, from satellites, and from weather balloons often contradict the projections of the computer simulations. For example, no warming of the lower troposphere (the atmosphere between 5,000 and 28,000 feet) has been recorded, even though climate simulators indicated that tropospheric warming should have already occurred due to increased carbon dioxide levels. Skeptical scientists contend that the fallibility of the IPCC's predictions raises serious doubts about the projected severity of global warming.

In June 2002, Bush publicly stated that humanity would be able to adjust to global warming. Some researchers are even more optimistic, maintaining that a warmer earth will result in lusher forests, increased food production, lower energy costs, and improved human health. Many scientists and environmentalists, however, fear that unchecked global warming will lead to dramatic increases in extreme weather that could uproot regional populations, augment the spread of infectious diseases, and disrupt worldwide economies. *At Issue: Is Global Warming a Threat?* examines the continuing controversy over the relationship between greenhouse gases and climate change.

1

Global Warming: An Overview

Kenneth Green

Kenneth Green is director of the environmental program at Reason Public Policy Institute, a nonprofit research organization headquartered in Los Angeles, California.

Certain atmospheric gases—such as carbon dioxide, methane, ozone, and nitrous oxide—trap solar energy and keep the temperature of the earth warmer than it would be without such gases. This "greenhouse effect" is beneficial because it has allowed life to proliferate on the planet. However, the levels of some greenhouse gases have increased since the late 1700s, causing a rise in the average global temperature. Some scientists maintain that human industrial activities are elevating concentrations of heat-trapping gases, and that the resultant global warming could lead to catastrophic climatic change. Other researchers dispute the theory that global warming is caused by human activity and contend that there is no evidence that the world's climate is undergoing dramatic change.

Popularly known as global warming, the theory of climate change has aroused intense controversy regarding the extent to which human activities affect global temperatures and weather patterns.

On the second day of the summer of 1988, when temperatures in Washington, D.C., had already reached a blistering 101 degrees Fahrenheit, James Hansen, director of NASA's Goddard Institute of Space Studies, offered an explanation that would earn him the title of "father of climate change." In a later recounting of that day's testimony before the U.S. Senate committee on energy and natural resources, Hansen noted: "I said three things. The first was that I believed Earth was getting warmer, and I could say that with 99 percent confidence. The second was that with a high degree of confidence we could associate the warming and the greenhouse effect. The third was that in our climate model, by the late 1980s and early '90s, there's already a noticeable increase in the frequency of drought."

Subsequent events in the summer of 1988 made Hansen's testimony seem prophetic. Even before he spoke, extreme temperatures afflicted the country. As early as May, record-high temperatures were set in 13 cities across the United States. Another 69 set records in June; 37 more in July; and 31 in August. In Los Angeles, the temperature hit 110 degrees in September, causing 400 electrical transformers to blow out in a single day.

The summer of 1988 was also unusually dry. The U.S. interior had not been so parched for 50 years. The Southeast had not been so dry for a century. Crop yields were down 30–40 percent across the nation as a whole, and boats were stuck in the mud of a shrunken Mississippi River. And it got worse. In August, massive fires in Yellowstone National Park burned nearly one million acres. In September, Hurricane Gilbert wreaked massive damage in the Yucatan, Mexico, and Texas. Almost 80 percent of Bangladesh was inundated by a flood—the fifth severe flood since 1980. The hottest year then recorded, 1988 was the capstone to four years of above-average warmth, making the 1980s the warmest decade in 127 years.

An old theory gains new life

With Hansen's prognosis and the global heat prostration of the 1980s, an old theory gained new life, picked up a few new layers, and was firmly linked to contemporary weather events. The core concept—called the greenhouse effect—was formulated in the 1800s, when scientists learned that certain gases in the atmosphere could function in a manner similar to the glass of a greenhouse, trapping solar energy and keeping the temperature of Earth's oceans, land surfaces, and atmosphere warmer than they would be otherwise. On that basis, the theory came to be known as global warming.

The greenhouse effect is fortunate for humanity, because in the absence of heat-trapping gases, Earth's average temperature would be about –2 degrees instead of the current average of 59 degrees, and our very existence would be impossible. In 1896, Swedish scientist Svante Arrhenius went further to suggest that the climate can be warmed by human activities involving the emission of carbon dioxide, a greenhouse gas.

If [some] scientists are correct, the secondhand impacts of global warming predicted by computer models—including sea-level rise, greater storm intensity, shifts in temperate zones, and so on— could be severe.

Hansen and other climate researchers added another layer to the theory of global warming, tying Earth's average temperature to other aspects of climate. Within this framework, which scientists now call climate change theory, they indicated that the average temperature of Earth's atmosphere could drive other climatic phenomena, such as rainfall, snowfall, glacier shrinkage, and so on. Moreover, they suggested that human activities have raised the atmospheric concentrations of greenhouse gases, leading not only to the heat of the late 1900s but also to floods, droughts, and any number of other unusual weather patterns since mankind started burning large quantities of fossil fuels.

Clearly, the concentrations of several greenhouse gases have risen dramatically. From the late 1700s to the present, carbon dioxide levels have increased by nearly 30 percent. Concentrations of methane, an even stronger warming gas, have increased nearly 150 percent since the beginning of the nineteenth century, while nitrous oxide levels have increased by about 22 percent from the preindustrial era. Ozone, which can warm the climate when present at low altitudes, has been increasing in concentration in the Northern Hemisphere over the last three decades. Ozone concentrations in the Southern Hemisphere are uncertain, while at the poles, low-altitude ozone concentrations seem to have fallen since the mid 1980s. Though these gases have major natural sources, the fingerprint of human action is evident in the changing concentrations.

Besides the greenhouse gases, other chemicals play a role (albeit poorly defined) in regulating global temperatures. Aerosols, a class of pollutants in the air, are liquid or solid particles small enough to remain suspended in air. Some aerosol particles reflect light or cause clouds to brighten, exerting a cooling effect on the atmosphere. Other aerosol particles are darker, absorb light, and can exert a warming effect.

Water vapor, an important component of the atmosphere, plays a major role in determining heat retention. But there is considerable controversy about the impacts of human activity on atmospheric water vapor content. Some studies point to an increase in atmospheric moistness and others to atmospheric drying over the last century.

Is global warming harmful?

Few dispute the fact that Earth's surface has warmed a bit in recent years. Fewer still dispute that human actions lead to greenhouse gas emissions. But neither of these issues is the most consequential. Rather, the important questions are, will bad things happen, and are humans responsible?

A slight increase in greenhouse gases and average global temperatures would generate only modest problems and might even have significant benefits for agriculture. But if Hansen and like-minded scientists are correct, the secondhand impacts of global warming predicted by computer models—including sea-level rise, greater storm intensity, shifts in temperate zones, and so on—could be severe.

Such computer models have been criticized for suffering from innumerable limitations. And the fact that politicians appear at storms, not at computer-modeling seminars, suggests that people's interpretation of the evidence of their own senses will decide the issue. Thus, for many, the central question is this: Do climate variations in the real world match up with what the theory predicts based on alterations in greenhouse gas concentrations? Let us examine some of the evidence.

Surface temperature measurements. Daily temperatures around the world have been recorded for about 150 years. About 2,000 ground-based weather stations feed daily measurements to this pool of evidence. Using these records for their 1995 report (still considered by many to be the "bible" of climate change theory), researchers with the Intergovernmental Panel on Climate Change (IPCC) noted an average warming trend, ranging from 0.5 to 1 degree since the year 1850. One-third to one-half of this warming has occurred since the mid 1970s.

This finding is basically validated by studies of tree-ring patterns and the contents of air bubbles trapped in soil or ice cores. Borehole records suggest that Earth's average temperature has increased by about 1.8 degrees over the last 500 years, while tree-ring studies in the Northern Hemisphere show that the twentieth century was somewhat warmer than average.

But evidence of surface temperature changes is not always clear cut. Tree-ring studies in the Southern Hemisphere, for example, have given contradictory results. In some places, tree rings reflect warmer-than-average temperatures for the twentieth century, while in other places they suggest only average temperatures.

Though global warming has theoretically been accelerating, there is little evidence that sea-level rise has accelerated accordingly.

Some climate researchers, such as Arizona State University's Robert Balling Jr., dispute the meaning of all or parts of this body of evidence. They point out that the record is too short to know whether the temperature increase over the last 100 years is part of a long-term trend or just a short-term spike. They note that such spikes have occurred naturally during Earth's temperature cycles, from the ice ages to warm periods. Conceding this point, the IPCC observes, "The recent [twentieth-century] warming needs to be considered in the light of evidence that rapid climatic changes can occur naturally in the climate."

In addition, the record may potentially be biased because most of the evidence comes from cities, where temperatures may be as much as 6–8 degrees higher than the surrounding countryside. This problem is based on the heat-retaining ability of dark surfacing materials in roads and buildings and the reduction in tree cover that often occurs with development.

Weather balloons and satellites. Since the 1960s, atmospheric temperatures have been measured by thermometers sent up with weather balloons. Equipped with assorted thermometers, balloons were flown to differing heights at various times. In addition, since the 1970s, satellites have been employed to measure Earth's temperature. The latter approach, involving special cameras that measure the heat given off by the upper atmosphere, afforded the first opportunity to actually measure the atmospheric temperature over Earth's entire surface.

But even the most modern methods raise questions. For instance, the decay of satellite orbits can bias the measurements. In addition, temperatures taken from balloons and satellites represent a much shorter record than surface temperature readings, and there is controversy over what they mean.

Weather balloon and satellite data do not match up well with the surface temperature record. While surface data suggest that Earth is warming near the ground, satellite and balloon measurements suggest that little or no warming has occurred higher in the atmosphere. This does not mean that Earth's average temperature has not changed, but it would appear to cast doubt on the theories and computer models that predict how greenhouse gases produced by human activities will lead to atmospheric changes.

Examining other factors

Direct temperature measurements represent one domain in the studies on climate change. To shed additional light on the issue, scientists have been examining a variety of other factors, such as rainfall, sea levels, snowfall, ice masses, and storms.

Rainfall. According to IPCC investigators, there has been a slight (about 1 percent) increase in global rainfall during the twentieth century. But like the temperature trends noted above, rainfall has varied unevenly, both geographically and over time. Complicating the situation, the places receiving more or less rainfall do not necessarily match up with those experiencing warmer- or cooler-than-average temperatures. Rainfall records also suffer from limitations in accuracy and the number of areas covered.

Sea levels and surface waters. Climate change theory suggests that as global temperatures increase, sea levels will rise, based on the thermal expansion of water and the melting of glaciers, ice sheets, ice caps, and sea ice. Some recent studies indicate that sea levels have risen about 7 inches over the last 100 years, but estimates range from 4 to 10 inches. Also, though global warming has theoretically been accelerating, there is little evidence that sea-level rise has accelerated accordingly.

Global warming would also be expected to influence bodies of surface water, such as lakes and streams, through alterations in rainfall and snowfall patterns. But the IPCC's last published report (in 1995) gives no clear evidence of widespread changes in annual stream flows and peak discharges of rivers in the world. While lake and inland sea levels have fluctuated, IPCC researchers point out that local effects make it difficult to use lake levels to monitor climate variations.

Snow and ice. Snowfall, snow depth, and snow cover (the total area covered by snow) should also be affected by climate change, but studies examining variations in these aspects have given mixed results. Snow cover has declined recently, with a higher percentage of moisture in cold areas coming down as rain rather than snow. But while the area covered by snow in the Northern Hemisphere diminished by about 10 percent over the past 21 years, some areas have received more snowfall than usual.

There is much controversy about the theory that global climate change is driven strongly by human activities.

Ice masses include glaciers, ice caps, ice sheets, and sea ice. From the records of explorers, climate investigators know that many of the world's glaciers have shrunk over the last 100 years. Even so, the IPCC's climate investigators admit that "continuous, long-term measurements of the mass balances of glaciers and ice caps are very limited." They further note that while some evidence suggests shrinkage of the ice sheets of Greenland and the Antarctic, other data suggest that the sheets are growing. There is also evidence that they may be doing both: growing on top and shrinking at the edges. Finally, it appears that neither hemisphere has experienced a change in the area of ocean covered by floating sea ice (such as icebergs) since 1973, when satellite measurements began.

Stormy weather. Climate change theory suggests that a warmer Earth would have more frequent heat waves, cold snaps, tornadoes, thunderstorms, dust storms, and so on. But these predictions have not been borne out on a global scale. In their last published review of the total body of evidence, IPCC investigators concluded that "overall, there is no evidence that extreme weather events, or climate variability, have increased, in a global sense, through the twentieth century."

Humanity's guilt: evidence or assumption?

While the greenhouse effect is relatively undisputed, there is much controversy about the theory that global climate change is driven strongly by human activities. Studies jockey back and forth about key elements of man-made climate change nearly every month on the pages of leading science journals such as *Nature* and *Science*. When it comes to attributing observed warming to human activities, the 1995 IPCC report leaves the question unanswered: "Some scientists maintain that these uncertainties currently preclude any answer to the question posed above. Other scientists would and have claimed . . . that confident detection of a significant [human-induced] climate change has already occurred."

Further complications have arisen as additional climate-modifying factors have been identified. Some scientists, such as Harvard astrophysicist Sallie Baliunas, look upward for the source of observed warming. According to a reconstruction of solar output levels from 1600 to the present, the Sun has clearly been running hotter, increasing the earthbound energy that constitutes the main input for Earth's surface temperature. Some studies suggest that increased solar output could be responsible for half the 1 degree rise in temperature from 1900 through 1970 and for a third of the warming seen since 1970.

Other scientists look downward for the cause of climate change. It has been discovered that continental shelves contain an unusual form of methane, called methane hydrate, trapped by high pressure and low temperatures. Some researchers, such as James Kennett at the University of California–Santa Barbara, theorize that the release of methane from these deep ocean reservoirs might constitute the main regulator of climate, with carbon dioxide merely going along for the ride.

Additionally, as discussed earlier, aerosols play a murky role in regulating climate. Some models suggest that, on a global basis, the cooling effect of aerosols offsets 20 percent or more of the predicted warming from the combined greenhouse gases. Other models dispute the significance of this masking effect.

In a curious turn of events, the potential cooling effects of aerosols have brought the debate full circle. In a recent publication, Hansen suggested that the warming potential of carbon dioxide released from fossil fuel burning may be canceled out by aerosols released in the burning process, and he now suspects that other pollutants—such as dark particulates, ozone, and methane—might have caused the warming seen in the latter years of the twentieth century.

How, then, shall we deal with climate change? While the proposals are many, they fall into two broad categories, as manifested by arguments during the U.S. presidential election in the year 2000.

On one side of the debate, people such as Albert Gore argue that the scientific evidence clearly shows that greenhouse gases produced by human activities will warm the climate and produce disastrous changes. Proponents of this "looming disaster" view call for immediate implementation of global greenhouse-gas reduction treaties and regimes, such as the Kyoto Protocol of 1997. Under such a regime, greenhouse gas emissions would be lowered mainly by reductions in fuel use and also through technology acceleration.

The economic impact of such approaches is a major bone of contention, yet it is widely acknowledged that implementation of the Kyoto Protocol by itself would produce just a trivial reduction in greenhouse gas concentrations. Jerry Mahlman, director of the Geophysical Fluid Dynamics Laboratory at Princeton University, has been quoted as saying that "the best Kyoto can do is to produce a small decrease in the rate of increase." In his view, "It might take another 30 Kyotos over the next century" to cut global warming down to size.

In the opposite camp are those who, like George W. Bush, argue that while the prospect of severe climate change is certainly worrisome, the evidence is not yet sufficient to take restrictive actions of the sort required by the Kyoto Protocol. Proponents of this "watchful wariness" view call for "no regrets" approaches to climate policy. These approaches would combine additional research with steps to address current problems, such as urban air pollution, thereby providing a climate-control benefit should man-made greenhouse gases turn out to be a significant cause of global warming. Such approaches would find economically harmless ways to speed up technological development, improve fuel economy, develop alternative fuel sources, and use other measures to reduce the urban air pollutants that—as Hansen now agrees—could be the major causes of any man-made climate change registered to date.

In a way, the issues of global warming can be summed up as a battle of axioms: "A stitch in time saves nine!" shouts the looming-disaster camp. "Act in haste, repent at leisure," reply the no-regrets supporters. Who will turn out to be right? Researchers on the cutting edge suggest that information from satellites launched during the past 20 years, combined with computer breakthroughs in the coming decades, will give us the ability to answer today's questions definitively—in about 50 years.

2

Human-Induced Global Warming Is a Serious Problem

Robert T. Watson

Robert T. Watson is chair of the Intergovernmental Panel on Climate Change, chief scientist at the World Bank, and a former science adviser to the White House.

Scientific data reveals that human activity is largely responsible for the rise in the average global temperature during the past fifty years. Furthermore, global warming is changing the world's climate, as evidenced by rising sea levels, thinning Arctic ice, record numbers of hot days, and heavy rains. Researchers project that the earth's temperature will rise by several degrees Fahrenheit over the next century, resulting in an increased risk of droughts, floods, and other forms of extreme weather. Unless the nations of the world agree to significantly reduce greenhouse-gas emissions, the environment, the economy, and human health will be adversely affected.

D o scientists now have compelling evidence of global warming? The simple answer is: "Yes." We have strong evidence that the climate is changing and that human activities were the primary cause of the changes during the 20th century.

First, we have detected change. Global-average temperature is rising, precipitation patterns are changing, glaciers are retreating, sea levels are rising and Arctic sea ice is thinning. Second, we can attribute most of the observed warming of the last 50 years to human activities rather than to changes in solar radiation or other natural factors. Third, because human activities will continue to change the atmosphere's composition throughout the 21st century, global warming can be expected to continue. This will result in significant projected increases in global-average temperature, in the number of hot days, in heavy precipitation events and in higher sea level.

In 1997, representatives from more than 100 governments met in Ky-

From "Q: Do Scientists Have Compelling Evidence of Global Warming? Yes: Rising Sea Levels Worldwide and Retreating Arctic Glaciers Are Ominous Signs," by Robert T. Watson, *Insight on the News*, March 12, 2001. Copyright © 2001 by News World Communications, Inc. Reprinted with permission.

oto, Japan, and agreed that industrialized countries should decrease their emissions of greenhouse gases. This decision was based in large part on the conclusions in the 1995 Assessment Report of the Intergovernmental Panel on Climate Change (IPCC). This report presented a careful and objective analysis of all relevant scientific, technical and economic information. It was prepared and peer-reviewed by more than 2,000 experts in the appropriate fields of science from academia, government, industry and environmental organizations worldwide.

In 1995, as today, the overwhelming majority of governments and scientific experts recognized that while scientific uncertainties existed, strong scientific evidence demonstrated that human activities were changing Earth's climate and that further human-induced climate change was inevitable. Hence, scientists and governments alike recognized in 1995 and reaffirm today that the question is not whether climate will change in response to human activities, but rather where (regional patterns), when (the rate of change) and by how much (magnitude).

The scientific evidence also indicates that climate change will, in many parts of the world, adversely affect human health, ecological systems (particularly forests and coral reefs), and important socioeconomic sectors, including agriculture, forestry, fisheries, water resources and human settlements. Developing countries are the most vulnerable, primarily because a larger share of their economies are in climate-sensitive sectors, and they do not have the institutional and financial infrastructures to adapt to climate change.

Serious climate changes

Since 1995, confidence in the ability of models to project future climate change has increased. This is because of their demonstrated performance in simulating key features of the climate system across a range of space and time intervals, most particularly, their success in simulating the observed warming during the 20th century. Certainly, models cannot simulate all aspects of climate. For example, they still cannot account fully for the differences in the observed trends in surface and midtropospheric temperatures during the last two decades.

On Jan. 22, 2001, in Shanghai, China, the IPCC released its latest report on climate-change science. This report was prepared by more than 600 scientists, peer-reviewed by more than 300 expert reviewers and further peer-reviewed and approved by nearly 100 governments. . . . Both the 1995 and 2001 IPCC reports say that the Earth's climate is warming, that human activities are implicated in the observed warming and that the Earth will warm several degrees Fahrenheit during the next 100 years as compared with only about 1 degree during the last 100 years.

However, the recent report makes an even stronger case that human-induced climate change is a serious environmental issue. Specifically, the recent report concluded, "Globally the Earth's climate is warmer today than at any time during the last 140 years. Direct ocean- and land-surface temperature measurements indicate that the global-average surface temperature has increased by 1.1 percent, or 0.4 F, since about 1860, with nighttime minimum temperatures warming at twice the rate of daytime maximum temperatures, with the land areas warming at twice the rate of the oceans, and with the 1990s being the warmest decade."

The temperature increase in the Northern Hemisphere in the 20th century was greater than for any other century in the last 1,000 years. A new analysis from direct and indirect data (including tree-ring and coral-reef records) shows the temperature increase in the 20th century is greater than that for any other century during the last 1000 years and that temperatures now are warmer than at any time during this period. This further demonstrates that the climate of the 20th century was unique. There is new and stronger evidence that most of the observed warming during the last 50 years is attributable to human activities, primarily the use of fossil fuels and deforestation. This statement, which is based on recent theoretical modeling and new data-analysis techniques, is much stronger than the pioneering finding made in the 1995 report that "the balance of evidence suggests a discernible human influence on global climate."

Climate change will, in many parts of the world, adversely affect human health, ecological systems . . . and important socioeconomic sectors, including agriculture, forestry, fisheries, water resources and human settlements.

A comparison of observed changes in temperature with simulations from several complex global-climate models shows that 20th-century climate changes cannot be explained by internal variability and natural phenomena. Simulations of climatic effects of changes in solar radiation and volcanic eruptions indicate that these natural phenomena may have contributed to the observed warming in the first half of the 20th century. But they cannot explain the warming in the latter half of the 20th century. In contrast, simulations that account for the impacts of both natural phenomena and human activities can account for both the time sequence and large-scale geographic patterns in surface temperature, as well as the trend in global-mean temperature.

Many observed changes in the Earth's climate are consistent with global-scale warming attributable to human influences. These include changes in precipitation, increases in sea level and oceanic temperatures, shrinking mountain glaciers, decreasing snow and ice cover and thinning Arctic sea ice.

Other predictions

The atmospheric concentrations of carbon dioxide and most other greenhouse gases are projected to increase significantly during the next 100 years. Emissions of greenhouse gases and sulfur dioxide will depend on a number of factors, including changes in population, economic growth and technological changes. But total cumulative carbon-dioxide emissions from all sources between 1990 and 2100 are projected to increase—in the absence of international action to address climate change—to a level of 770 to 2,540 billion tons of carbon. In contrast, total cumulative emissions between 1800 and the present have been about 400 billion tons of carbon. Despite the uncertainty in future emissions, all plausible projections suggest a significant increase in the atmospheric concentration of carbon dioxide, rising to a level of 540 to 970 parts per million (ppm) by 2100. Compare this with the preindustrial concentration of 280 ppm and the current concentration of about 368 ppm.

The temperature increase in the Northern Hemisphere in the 20th century was greater than for any other century in the last 1,000 years.

The atmospheric concentrations of sulfur dioxide are, in most cases, projected to decrease during the next 100 years. Sulfur-dioxide emissions, which tend to cool the atmosphere, are projected to range from about 11 million to 93 million tons per year, in contrast with the 1990 emissions ceiling of about 70 million tons. In general, these projections are much lower than those in 1992 because most countries likely will try to reduce sulfate acid deposition. These lower projections enhance the magnitude of climate change by reducing the sulfate-aerosol cooling effect.

Globally, average surface-air temperatures are projected to rise 2.5 F to 10.4 F between 1990 and 2100, with most land areas warming more than the global average by up to 40 percent. A simple climate model simulating the response of seven complex global-climate models projects globally averaged surface-temperature increases of 2.5 F to 10.4 F between 1990 and 2100 for the full range of plausible trends in greenhouse gas and sulfur-dioxide emissions. This range of projected temperature change between 1990 and 2100 is higher, by 1.8 F to 6.3 F, than that reported in the assessment report, primarily because of the lowering of projected sulfur-dioxide emissions.

Globally averaged precipitation is projected to increase, but both increases and decreases will occur depending on the specific global region. Warmer temperatures will enhance evaporation and precipitation by a few percent under all plausible emissions scenarios. Precipitation is projected to increase in both summer and winter over high-latitude regions; in winter over northern mid-latitudes, tropical Africa and Antarctica; and in summer over South and East Asia. Conversely, precipitation is projected to decrease in winter over Australia, Central America and Southern Africa. However, a key finding is that more-intense precipitation events are very likely over most regions of the world, consistent with changes already measured during the 20th century in northern mid and high latitudes. In addition, an increased risk of droughts and floods associated with the El Nino phenomena is very likely, even if there is no change in the intensity of the El Nino phenomena.

Global sea level is projected to rise by about 4 inches to 35 inches between 1990 and 2100. The top end of this range is quite close to the range projected in the 1995 IPCC assessment despite the higher temperature projections, primarily because of lower estimated contributions from Greenland and Antarctica.

In conclusion, the recent IPCC report provides strong evidence for human influence on the Earth's climate system. The good news is that the majority of experts believe significant reductions in net greenhouse-gas emissions are technically feasible. This is because of an extensive array of cost-effective technologies and policy measures in the energy supply, energy demand, agricultural and forestry sectors. In particular, the cost of reducing carbon-dioxide emissions for developing countries significantly can be lowered by industrialized countries purchasing "carbon credits" from developing countries which use the monies for abatement of carbon-dioxide emission. The bad news is that not only the climate is changing because of human activities, but evidence is mounting that ecosystems are as well. Given the fact of compelling scientific evidence, it is time for nations to determine how best to respond.

3

Human-Induced Global Warming Is Insignificant

Sallie Baliunas

Sallie Baliunas is an astrophysicist at the Harvard-Smithsonian Center for Astrophysics and deputy director of the Mount Wilson Observatory. She is also chair of the Science Advisory Board at the George C. Marshall Institute in Washington, D.C.

Humanity's use of fossil fuels, which produces heat-absorbing gases such as carbon dioxide and methane, does not significantly contribute to global warming. Most of the global warming during the twentieth century occurred before 1940, even though the levels of human-produced carbon dioxide increased greatly after 1940. Moreover, climate forecasts drawn from computer simulations—which tend to predict an acceleration of global warming for the future—are highly unreliable. For example, no warming of the lower troposphere (the atmosphere between 5,000 and 28,000 feet) has been recorded, even though climate simulators indicated that tropospheric warming should have already occurred due to increased carbon dioxide levels. Since human-induced global warming is minor, there is no need to sharply curtail the use of fossil fuels—currently the most accessible and cost-effective form of energy.

Editor's Note: The following viewpoint is abridged from a speech delivered at Hillsdale College on February 5, 2002, at a seminar cosponsored by the Center for Constructive Alternatives and the Ludwig von Mises Lecture Series.

The evolution from fire to fossil fuels to nuclear energy is a path of improved human health and welfare arising from efficient and effective access to energy. One trade-off is that energy use by human beings has always produced environmental change. For example, it has resulted in human artifacts marking the landscape, the removal of trees from major areas for wood burning, and region-wide noxious air pollution from coal

From "The Kyoto Protocol and Global Warming," by Sallie Baliunas, *Imprimus*, the National Speech Digest of Hillsdale College (www.hillsdale.edu), March 2002. Copyright © 2002 by *Imprimus*. Reprinted with permission.

burning. On the other hand, ready availability of energy that produces wealth through the free market system provides ways to remedy or minimize environmental damage from energy use.

With widespread industrialization, human use of coal, oil and natural gas has become the centerpiece in an international debate over a global environmental impact, viz., global warming. Fossil fuels provide roughly 84 percent of the energy consumed in the United States and 80 percent of the energy produced worldwide. An attempt to address the risk of deleterious global warming from the use of these carbon dioxide–mitting fuels is embodied in the Kyoto Protocol and its attendant series of international negotiations. But on scientific, economic and political grounds, the Kyoto Protocol as an attempt to control this risk while improving the human condition is flawed.

What would Kyoto do?

Projections of future energy use, applied to the most advanced computer simulations of climate, have yielded wide-ranging forecasts of future warming from a continued increase of carbon dioxide concentration in the air. The middle range forecast of the estimates of the United Nations Intergovernmental Panel on Climate Change, based on expected growth in fossil fuel use without any curbs, consists of a one degree Celsius increase over the next half century. A climate simulation *including* the effect of implementing the Kyoto Protocol—negotiated in 1997 and calling for a worldwide five percent cut in carbon dioxide emissions from 1990 levels—would reduce that increase to 0.94 degree Celsius. This amounts to an insignificant 0.06 degree Celsius averted temperature increase. [See Chart 1. The jagged line tracks the forecast of increasing temperatures through 2050, based on the Hadley Center's model. The upper straight

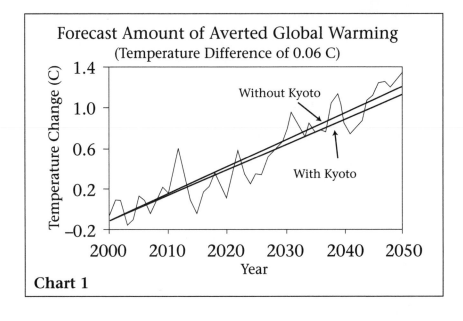

line is the linear trend fit to the model's forecast temperature rise without implementation of Kyoto, and the lower straight line is the linear trend *with* implementation.]

To achieve the carbon dioxide emission cuts by 2012 that are required under the Kyoto agreement, the United States would have to slash its projected 2012 energy use by about 25 percent. Why, then, are the temperature forecasts so minimal in terms of averted global warming? The answer is that countries like China, India and Mexico are exempt from making emission cuts, and China alone will become the world's leading emitter of carbon dioxide in just a few years.

Most economic studies indicate that the cost of the Kyoto carbon dioxide emission cuts to the U.S. would amount to between $100 billion and $400 billion per year. One major reason these costs are so high is that past U.S. energy policy has been constrained by political influences. For example, substantially expanding the number of U.S. nuclear power plants and reducing the number of coal plants would enable the U.S. to meet both its future energy needs and Kyoto's mandated carbon dioxide emission reductions. But no nuclear power plants have been built in the U.S. in over 20 years, owing to non-technical factors.

No catastrophic human-made global warming effects can be found in the best measurements of climate that we presently have.

Over the same period, renewable energy sources like wind and solar power have been discussed to the point of distraction. But these are boutique energy sources: they produce relatively minute amounts of energy and do so intermittently. While they may be cost-effective in limited locales, they are unreliable for large-scale electricity generation. (As a side note, often overlooked is the enormous environmental footprint that wind and solar farms would require. For example, to replace a conventional 1000 megawatt coal plant that spans tens of acres would require an isolated, uninhabited area with correct meteorological conditions of roughly 400 square miles on which to place over 2,000 wind turbines, not to mention the associated imprint of high-power transmission lines, roads, etc. Solar panel farms would produce environmental blight and degradation over a similarly sized landscape.)

The Kyoto Protocol also has the potential to worsen international relations. The struggling economies of the world rely on the U.S. to maintain stability and to provide aid and economic opportunity as a trading partner. While the developing nations are exempt from making carbon dioxide emission cuts, the severe economic impact on the U.S. would dramatically curtail its ability to continue to promote international stability and to help improve those nations' economies.

What does science say?

Whereas the economic catastrophe that would occur as a result of implementing the Kyoto Protocol is a certainty, the likelihood of an environmental catastrophe resulting from a failure to implement Kyoto is extremely speculative.

The facts in scientific agreement concerning global warming are as follows:

- As a result of the human use of coal, oil and natural gas, the air's carbon dioxide content (along with the content of other human-produced greenhouse gases like methane) is increasing.
- The greenhouse gases absorb infrared radiation and, as a result, should retain some energy near the surface of the earth that would otherwise escape to space.
- Based on current ideas about how climate should work, the surface temperature should warm in response to the addition of the small amount of energy arising from a benchmark doubling of the air's carbon dioxide content.
- The main greenhouse effect is natural and is caused by water vapor and clouds. But the impacts of these greenhouse factors are for now greatly uncertain. In other words, the reliability of even the most sophisticated computer simulations of the climate impacts of increased carbon dioxide concentration rests heavily on the use of factors that science does not understand. To put this in perspective, the uncertainties surrounding the use of clouds and water vapor in climate simulations—not to mention other important factors like sea-ice changes—are at least ten times greater than the effect of the variable being tracked, i.e., the temperature rise caused by increased carbon dioxide levels in the air.
- Finally, in the absence of any counterpoising or magnifying responses in the climate system, the global average rise in temperature is roughly one degree Celsius or less at equilibrium for a *doubling* of the air's carbon dioxide concentration. That is meager warming for so profound a change in the air's carbon dioxide content. Indeed, it is within the range of climate's natural variability.

One key question in the debate over global warming is the following: What has been the response of the climate thus far to the small amount of energy added by humans from increased carbon dioxide in the air? This question is important because, in order to prove the reliability of future climate forecasts from computer simulations, those simulations need to prove that they are reliable at explaining past temperature change. They have not yet done so.

Plants have flourished . . . due directly to the fertilization effect of increased carbon dioxide in the air.

In the twentieth century, the global average surface temperature rose about 0.5 degrees Celsius. At first glance, one might think this attributable to human fossil fuel use, which increased sharply over the past 100 years. But a closer look at twentieth century temperatures shows three distinct trends: First, a strong warming trend of about 0.5 degrees Celsius began in the late nineteenth century and peaked around 1940. Then, oddly,

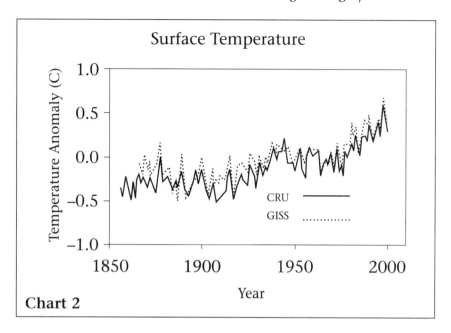

Chart 2

there was a cooling trend from 1940 until the late 1970s. And a modest warming trend occurred from the late 1970s to the present. [See Chart 2, illustrating surface temperature changes sampled worldwide and analyzed by Cambridge Research Unit (solid line) and NASA-Goddard Institute of Space Studies (dotted line). Both lines show these three distinct phases.]

How do we interpret this data? We know that about 80 percent of the carbon dioxide from human activities was added to the air *after* 1940. Thus increased carbon dioxide in the air cannot account for the pre-1940 warming trend. That trend had to be largely natural. Then, as the air's carbon dioxide content *increased* most rapidly, temperatures *dropped* for nearly 40 years. And it seems that human effects amount at most to about 0.1 degree Celsius per decade—the maximum increase in warming seen after the 1970s.

How, then, does the observed surface-warming trend in recent decades—even assuming it is all due to human activity—compare to the results of climate change computer simulations?

Looking back at Chart 1, climate simulations predict that a smooth, linear rise of at least twice the observed trend should already be occurring, and that it will continue through the next century. Given that the warming trend has been observed to be at most 0.1 degree Celsius per decade from human activities, these future forecasts appear greatly to exaggerate the future warming and should be adjusted downward to, at most, one degree Celsius warming by 2100. That amount of warming would be very similar to natural variability, which humans have dealt with for thousands of years. Indeed, it would likely return climate conditions to those experienced in the early centuries of the second millennium, when widespread warming is indicated by numerous proxies of climate, such as glaciers, pollen deposits, boreholes, ice cores, coral, tree growth, and sea and lake floor sediments. (It is interesting to note that this so-called Medieval Climate Optimum is associated with the settling of Greenland and Iceland, travel by the Vikings to Newfoundland, higher crop yields and generally rising life spans.)

New data

In addition to what we can deduce from surface temperature data, U.S. leadership in new space instruments and in the funding of global research has yielded atmospheric temperature data that also indicates a lesser human-made global warming trend than is forecast by climate simulations.

According to these simulations, a readily detectable warming of the lower troposphere (roughly 5,000 to 28,000 feet altitude) must occur with the presence of increased atmospheric carbon dioxide concentration. But records from NASA's microwave sounder units aboard satellites show no such trend. These satellite records are essentially global, in contrast with records of surface temperatures, which cover a mere fifth of the planet. And what emerges from them is that while the tropospheric temperature does vary over short periods—for example, with the strong El Niño warming pulse of 1997 and 1998—no meaningful warming trend is observed over the 21-year span of the record. [See Chart 3, illustrating monthly averaged temperatures for the lower troposphere from instruments onboard NASA satellites. Even taking into account the 1997–98 El Niño event, the linear trend is only +0.04 degree Celsius per decade. Data are from http://wwwghcc.msfc.nasa.gov/temperature/.]

It should be noted in passing that there has been a proposed explanation for the lack of a significant human-made warming trend in the

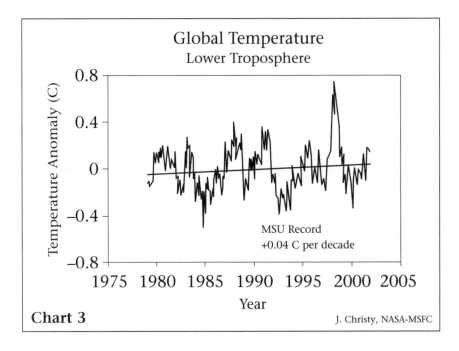

Global Temperature
Lower Troposphere

MSU Record
+0.04 C per decade

Chart 3

J. Christy, NASA-MSFC

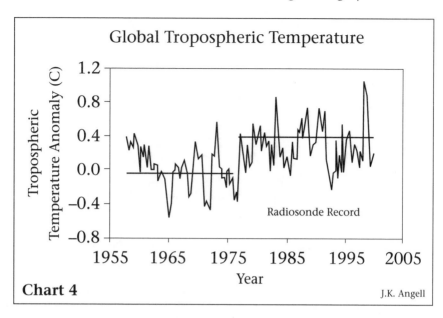

Global Tropospheric Temperature

Chart 4

J.K. Angell

lower troposphere. This explanation contends that human-induced global warming is masked because of soot from sulfur dioxide and other human-made aerosols, which cool the atmosphere. But this idea of a widespread aerosol shading effect fails the test by the scientific method, because the Southern Hemisphere—which shows no long-term warming trend at all—is relatively free of aerosols.

In addition to satellite records, we have a radiosonde record from balloons that goes back over four decades. This record obviously lacks the dense spatial coverage of satellite measurements. Nevertheless, it too shows no warming trend in global average temperature that can be attributed to human effects. It records the strong warming in 1976–77 known as the Great Pacific Climate Shift, resulting from a natural, periodic shift in the Pacific—the Pacific Decadal Oscillation—which is so significant that global average temperatures are affected. [See Chart 4, which illustrates the seasonal average temperature anomaly sampled worldwide for the lower troposphere as measured by radiosonde instruments carried aboard balloons. Although a linear trend of +0.09 degree Celsius per decade is present if fitted across the entire period of the record, the trends before and after the abrupt warming of 1976–1977 (straight horizontal lines) indicate no evidence of significant human-made warming. Data are from http://cdiac.esd.ornl.gov/ftp/trends/temp/angell/glob.dat.] Furthermore, the Pacific now seems to have shifted, perhaps in 1998–99, back to its pre-1976 phase, which should produce cooler temperatures, especially in Alaska and in the global average.

Thus according to our most reliable data, when compared to the actual measurements of temperature over the past four decades, computer simulations overestimate to some degree the warming at the surface and decidedly exaggerate warming in the lower troposphere. And given that the models have overestimated past warming trends, they presumably also exaggerate the warming to be expected in the future. This inaccuracy is not surprising. Computer simulations of climate must track over five million parameters relevant to the climate system. To simulate climate change for a period of several decades is a computational task that requires 10,000,000,000,000,000,000 degrees of freedom. And to repeat, such simulations require accurate information on two major natural greenhouse gas factors—water vapor and clouds—whose effects we do not yet understand.

Human-made global warming is relatively minor and will be slow to develop, affording us an opportunity to continue to improve observations.

Finally, it should be mentioned that in looking for natural factors influencing the climate, a new area of research centers on the effects of the sun. Twentieth century temperature changes show a strong correlation with the sun's changing energy output. Although the causes of the sun's changing particle, magnetic and energy outputs are uncertain—as are the responses of the climate to solar changes—the correlation is pronounced. It explains especially well the early twentieth century temperature increase, which, as we have seen, could not have had much human contribution. [See Chart 5, illustrating the change over four centuries of the Sunspot Number, which is representative of the surface area coverage of the sun by strong magnetic fields. The low magnetism of the seventeenth century, a period called the Maunder Minimum, coincides with the coldest century of the last millennium, and there is sustained high magnetism in the latter twentieth century. See also Chart 6, showing that changes in the sun's magnetism—as evidenced by the changing length of

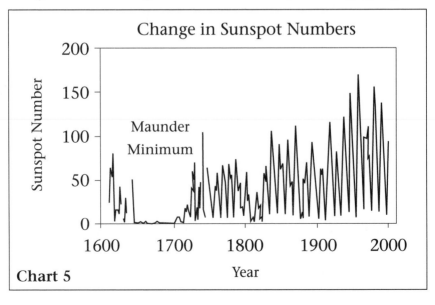

Chart 5

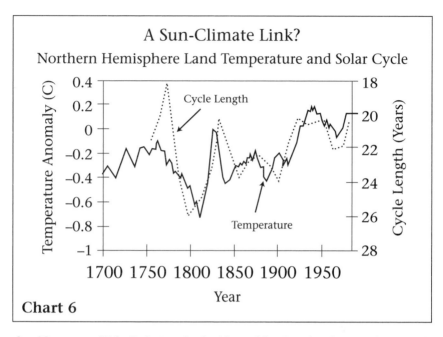

A Sun-Climate Link?
Northern Hemisphere Land Temperature and Solar Cycle

Chart 6

the 22-year or Hale Polarity Cycle (dotted line)—closely correlates with changes in Northern Hemisphere land temperature (solid line). The sun's shorter magnetic cycles are more intense, suggesting a brighter sun during longer cycles. Lags or leads between the two curves that are shorter than 20 years are not significant, owing to the 22-year time frame of the proxy of brightness change. In this chart, the record of reconstructed Northern Hemisphere land temperature substitutes for global temperature, which is unavailable back to 1700.]

Two conclusions

Two conclusions can be drawn about global warming and human energy use:

- No catastrophic human-made global warming effects can be found in the best measurements of climate that we presently have.
- The longevity, health, welfare and productivity of humans have improved with the use of fossil fuels for energy, and the resulting human wealth has helped produce environmental improvements beneficial to health as well.

In light of some of the hysterical language surrounding the issue of greenhouse gases, it is also worth noting that carbon dioxide, the primary greenhouse gas produced by burning fossil fuels, is not a toxic pollutant. To the contrary, it is essential to life on earth. And plants have flourished—agricultural experts estimate a ten percent increase in crop growth in recent decades—due directly to the fertilization effect of increased carbon dioxide in the air.

It is good news, not bad, that the best current science offers little justification for the rapid cuts in carbon dioxide mandated by the Kyoto Protocol. This science indicates that human-made global warming is relatively minor and will be slow to develop, affording us an opportunity to continue to improve observations and computer simulations of climate. These will serve to better define the magnitude of human-made warming, and allow development of an effective and cost-effective response.

Given this science, what is impelling the Kyoto Protocol's international momentum? One strong factor is the "Precautionary Principle" in environmental regulation. This principle disallows an action that might harm the environment until the action is certain to be environmentally harmless. It is antithetical to science in practice, because it sets the impossible goal of proving harmlessness with certainty. In addition, a policy of "doing something" is promoted as "insurance" against possible risk to the earth. This idea of insurance as a prudent hedge is wrong on two counts, notwithstanding the lack of scientific evidence of significant human-made warming. First, the actuarial notion of insurance is that of a carefully calculated premium, paid against a reasonably well-known risk in outcome and probability of outcome. But in the case of human-made global effects, the risk, premium and outcomes cannot be well defined. Second, the notion that implementing the Kyoto Protocol is effective insurance ignores the fact that the actual averted warming that would result is inconsequential. Indeed, the underlying basis for current international negotiations is the Rio Treaty of 1992, which specifically states that concentrations of greenhouse gases in the atmosphere, not emissions, be stabilized. In order to stabilize the air's *concentration* of greenhouse gases, emissions would have to be cut some 60 to 80 percent.

For the next several decades, fossil fuels are key to maintaining Americans' way of life and improving the human condition. According to the scientific facts as we know them today, there is no environmental reason we should not continue using them.

4

Global Warming Is Not a Serious Problem

Kevin A. Shapiro

Kevin A. Shapiro is a researcher in neuroscience at Harvard University.

There is not enough evidence to support the theory that humans are responsible for global warming or that this warming will lead to catastrophic changes in climate. The higher incidence of severe weather in the Pacific during the past decade is not necessarily the result of global warming, and recent climate-prediction research shows no correlation between rising global temperatures and an increase in the frequency of storms. During the twentieth century, the average world temperature increased by only one-half of a degree centigrade. Although some climatologists predicted, in 1980, that this warming would cause the polar icecaps to melt and sea levels to rise drastically, these forecasts proved to be wrong. The United States should not cave in to international pressure to stringently reduce its output of greenhouse gases—such a reduction would be economically unfeasible and scientifically unsound.

Natives of Hawaii, inured by more than a thousand years of island life to the vagaries of the weather and the seas, have a somewhat elliptical saying: "the mists are those that know of a storm upon the water." It can be taken to mean that those nearest to something are the first to become aware of what is happening to it. Using similar reasoning, perhaps, many environmentalists today regard the small islands that dot the Pacific as a sort of planetary weathervane, outcrops of flora and fauna that are sensitive indicators of large-scale shifts in the ecological balance of the earth. If these islands are already beginning to buckle under the stresses imposed on the planet by human activity, it is a sign that we must act quickly lest catastrophe result.

An alarming presentation of this argument can be found in *Rising Waters: Global Warming and the Fate of the Pacific Islands,* an hour-long documentary that aired on PBS in April 2001 on Earth Day. *Rising Waters* paints a picture of island nations on the veritable brink of ruin: homes destroyed in the wake of storms or threatened by eroding shorelines, churchyards and cropfields inundated by the rising sea, and shoals of once-vivid coral bleached by overheated waters. On camera, fishermen complain of poor hauls; a Samoan environmentalist laments the looming disappearance of his cultural heritage; Teburoro Tito, the president of tiny Kiribari, speaks glumly of the possibility that the entire populace of his cluster of atolls will have to be relocated.

What is causing this potentially immense disruption? *Rising Waters* mentions several factors, including seasonal weather fluctuations and overdevelopment, but ultimately it places most of the blame on a long chain of processes at the end of which is: global warming. The nature of this menace is well known and has been widely discussed. Increases in the industrial emission of gases like carbon dioxide (CO_2), it is said, have caused the atmosphere to absorb infrared radiation that would otherwise be reflected back into outer space. The resulting "greenhouse effect" lifts the average temperature of the earth's surface. Among the many consequences are rising sea levels caused by the melting of the polar ice caps and increases in the frequency and intensity of storm activity.

Model [weather] projections suggest no clear trend, so it is not possible to state whether the frequency, intensity, or distribution of tropical storms and cyclones will change.

Though *Rising Waters* offers the disclaimer that the earth's climate is a complex and somewhat unpredictable system—"we don't know how it behaves completely," says Fred MacKenzie, a professor of oceanography at the University of Hawaii—its overarching message is that unregulated CO_2 emissions have already begun to heat the planet to dangerous levels. To forestall further warming, we must cut those emissions globally by as much as 80 percent over the next several decades. Alas, as *Rising Waters* notes with a hint of impending doom, the prospects for such a cut are not auspicious.

On this last point, the documentary is certainly correct. Talks in the Hague on implementing the 1997 Kyoto Protocol, an international agreement aimed at reducing the CO_2 emissions of industrial nations to pre-1990 levels by the year 2012, collapsed in December 2000, in the last month of Bill Clinton's presidency. By mid-March 2001, the Bush administration had announced it would not seek to regulate the CO_2 emissions of power plants, provoking an outcry from environmentalists and angering European leaders who maintain (in the words of Dutch prime minister Wim Kok) that the United States is acting "irresponsibly." Two weeks later, President George W. Bush declared that it made "no sense" for the United States to pursue implementation of the Kyoto Protocol. European governments, positively livid, dispatched an emergency delegation to Washington, but to no avail; they now plan to assemble an international coalition aimed at "shaming" the United States into reconsidering its stance. Another round of talks on Kyoto will be held in Bonn in July 2001, and the conflict over global warming is certain to deepen in the months and years ahead.

Defying common sense

Against this backdrop, *Rising Waters* can only serve to underscore the now almost incessant warnings about the disaster that awaits us if we fail to change our profligate energy habits. Global warming has already been blamed for ecological hazards ranging in scale from disruptions in the migration patterns of butterflies and declining amphibian populations to extreme weather events, droughts, and food shortages in farflung portions of the globe. And the dangers that lie ahead are said to be far worse, if not horrific: famine brought on by widespread agricultural failure, an increase in epidemics of infectious disease, even mass extinctions of animal and human populations.

If anything remotely resembling this scenario is likely, it is not hard to see why so many Europeans, and with them many Americans, are apoplectic over President Bush's determination to scrap the Kyoto deal, the fruit of years of intense multinational discussions among lawmakers, economists, scientists, and environmentalists. Senator Joseph Lieberman has even promised a congressional investigation of the President's environmental decisions, declaring that they "ignore the public interest and defy common sense."

Is Lieberman right? There are indeed many things about the global-warming debate that "ignore the public interest and defy common sense." But the decision to abandon the Kyoto Protocol is not one of them.

In a sense, the decision was hardly even newsworthy. The agreement has been effectively dead—at least as far as the United States is concerned—since shortly after it was negotiated in 1997. For no sooner did Clinton's negotiators return from Japan than the Senate voted 95-0 to oppose ratification of any treaty that would impose significant burdens on our national economy and that lacked "specific scheduled commitments" for emissions reductions in what are now known as "developing" countries. As Kyoto has never been amended to address these concerns, it is perplexing that any could have continued to regard the accord as viable.

According to the latest data, the polar icecaps do not appear to be melting at all.

Indeed, far more inscrutable than President Bush's final rejection of Kyoto is the vast amount of rhetorical and diplomatic effort that has been and continues to be expended on the agreement's behalf. Even apart from the unanimous vote in the Senate, there are serious questions about whether the provisions of the treaty could ever be implemented and enforced, and therefore about whether it really represents a workable mechanism for managing climate change.

From its very inception, as the analyst David G. Victor shows in his monograph *The Collapse of the Kyoto Protocol and the Struggle to Slow Global Warming*, the Kyoto Protocol was a product of diplomatic wishful thinking. For one thing, the limits it called for on greenhouse gas emissions were draconian. Thus, by 2012 the United States would have been required to reduce CO_2 emissions to 7 percent below 1990 levels—a modest-sounding target until one considers that by the end of 1999, emissions were already 12 percent above 1990 levels and were continuing to rise. Compliance with Kyoto would therefore have required a likely cut of as much as 30 percent by the time the treaty took effect in 2008. Not only would this cost hundreds of billions of dollars in Gross Domestic Product (GDP) but, because most greenhouse gases are released in the course of burning fossil fuels for energy, cutbacks on such a scale would deal a major blow to significant sectors of the U.S economy—particularly electricity generation, which is already struggling mightily to keep pace with demand.

The agreement was also exceedingly inequitable. Russia, for example, would have been required only to freeze its emissions at 1990 levels; but because the Russian economy has contracted sharply since the collapse of the Soviet Union, its emissions are already far below target, and are unlikely to recover by 2008. Though it remains a significant industrial polluter, Moscow would thus be required to do absolutely nothing. South Korea and Mexico, now formally considered "developed" countries (as defined by membership in the Organization for Economic Cooperation and Development), have for their part also not agreed to curtail emissions.

At the same time, Kyoto sets no targets at all for the developing nations, though these countries will account for half the world's greenhouse gases by 2020. The two largest such nations, India and China, have refused outright to accept any limits on their emissions output.

Under the worst-case scenario . . . the oceans should rise no more than a foot over the next century.

In short, the Kyoto Protocol demands that the United States hobble its economy with drastic cuts in energy production, while Russia, India, China, and other nations enjoy the freedom to grow untrammeled. To deal with this gross imbalance, a number of observers have proposed amending the agreement. One proposal involves altering the way emissions are accounted for—for example, by permitting industrialized countries to earn "credits" if they maintain or create carbon sinks, i.e., forest and soil zones that absorb CO_2. Another alternative would be to allow trading, whereby industrialized countries could buy the right to emit carbon dioxide from those nations whose emissions are below targeted levels.

Both of these ideas have their attractions for the United States, but they also entail immense practical and political difficulties. On the positive side, the U.S. might offset its Kyoto obligations by counting carbon sinks that resulted from intentional changes in land-use policy. If, in addition, it were permissible to count those resulting from unintentional changes (like the spontaneous reforestation of abandoned agricultural lands), we might no longer be a net emitter. But an amendment of this sort would almost certainly prove unacceptable to Europe and Japan, which, unlike the U.S., have limited capacity to plant new forests. A more fundamental problem is that the Kyoto Protocol provides no standard definitions, methods, or data for quantifying the absorption of CO_2 by trees and soils, making it easy for nations to cheat by claiming credit for carbon sinks that are short-lived or even nonexistent.

Emissions trading is beset with its own difficulties. The present terms of the Kyoto Protocol would seem to award countries with low baselines—like Russia—a windfall in fictitious credits, the sale of which would result in no reduction in global emissions whatsoever. David Victor has correctly spelled out the political implications of any such arrangement: "No Western legislature will ratify a deal that merely enriches Russia and Ukraine while doing nothing to control emissions and slow global warming."

If the most widely discussed ways of amending the Kyoto agreement are infeasible, what then? Policy analysts like Victor continue to hold out hope that the Bush administration will develop a coherent approach to global warming—perhaps a modified trading system combined with international taxes on CO_2 emissions and supplemented by investments in new technology. As for the Bush administration, the President himself has spoken of global warming as a "serious problem," and the U.S. will be participating in the summer 2001 talks in Bonn with the hope of finding a workable alternative to Kyoto.

A scientific shell-game

The operative assumption here, of course, is that man-made climate change is a real phenomenon, and that averting catastrophe requires doing something about it, and soon. As this assumption has increasingly come to be taken for granted, disputing it has become commensurately perilous, especially for politicians. According to a 1997 poll taken for the World Wildlife Federation, two thirds of American voters regard global warming as a "serious threat" and support an international agreement to cut greenhouse-gas emissions, even if this comes at some economic cost. A full three quarters endorse the view that "the only scientists who do not believe global warming is happening are paid by big oil, coal, and gas companies to find the results that will protect business interests." Only 15 percent accept the statement that "scientists disagree among themselves" about the extent of the coming danger.

Clearly, climate change is no longer an issue up for grabs. Even if the public could be persuaded that the Kyoto Protocol would be disastrous for the U.S. economy and is the result of junk diplomacy, it would be far harder for a politician to make the case that the research behind Kyoto is junk science, too. But much of it is.

Let us return for a moment to those Pacific islands. It is undeniable that they have been buffeted by a series of severe storms in the past decade, accompanied by unusually intense episodes of the El Nino-Southern Oscillation (ENSO) phenomenon, a periodic fluctuation in sea temperature in the tropical Pacific that has been observed since the twentieth century. What is not clear is whether these have anything to do with global warming.

Storm activity in the Pacific varies from year to year; 1998 saw an above-average incidence of tropical storms, while 1997 was comparatively quiet. The cause of this variation remains unknown. The ENSO phenomenon is not well understood, and it is not predicted by any model of climate change. A United Nations body called the Intergovernmental Panel on Climate Change (IPCC) has rightly observed that while many small island states fear that "global warming will lead to changes in the

character and pattern of tropical cyclones (i.e., hurricanes and typhoons)," this fear is not confirmed by the most recent research. Rather,
"model projections suggest no clear trend, so it is not possible to state
whether the frequency, intensity, or distribution of tropical storms and
cyclones will change."

And what of rising waters? In 1980, climatologists predicted that
global warming would melt the polar icecaps, causing sea levels to rise
more than 25 feet over the course of the next century. Such an event
would undoubtedly be disastrous not only for the Pacific islands but also
for densely populated coastal regions in all parts of the world.

---

*At present there is no basis in scientific evidence for
. . . drastic action [on global warming].*

Fortunately for those of us in Boston, Miami, New York, and Los Angeles, the deluge failed even to begin to materialize. According to the latest data, the polar icecaps do not appear to be melting at all. The 2001
IPCC report discerns "no significant trends" in the extent of Antarctic
sea-ice since 1978, when reliable satellite measurements began to be
taken; nor, at the other pole, is there evidence from satellite records that
the air above the Arctic has warmed substantially.

With the polar caps essentially intact, it does not come as a surprise
that sea levels have risen only a paltry 2 millimeters per year in the mid-
1990's—roughly the same rate observed over the past 100 years. Even the
gloomiest doomsayers have been compelled to jettison the dire forecasts
put forward in 1980. Under the worst-case scenario now envisioned by
the IPCC, the oceans should rise no more than a foot over the next century; not nearly enough to pose a major threat. And this forecast is in turn
based on the assumption that sea levels will increase by approximately 5
millimeters per year, give or take 3 millimeters—in other words, the rate
of rise may not change at all.

As for the climate itself, despite the alarmed rhetoric from so many
quarters, we do not know for certain that it is even changing in significant ways. It is an established fact that the earth's climate has warmed
slightly over the past century. Average temperatures near the surface have
risen since 1900 and are now probably higher than they have been at any
time in the past 600 to 1,000 years. But that statement more or less ex-
hausts the scientific consensus. On every other important question—
what the major causes of global warming are, what its effects will be,
whether we should try to prevent it and, if so, how—there is considerable
uncertainty.

Most of what we "know" about the earth's future is derived from
enormously sophisticated computer models that utilize millions of parameters to simulate the earth's climate. These models are still far from reliable. The editors of *Nature,* arguably the world's most prestigious scientific journal, pointed out on March 15, 2001, that "the accuracy of any
model depends significantly on the quality of the underlying raw data."
But the quality of the data being used for climate prediction, they go on
to state, is "patchy." For example, it is not at all easy to measure the

amount of sunlight absorbed by the atmosphere or reflected by its surface back into space—and yet this one key parameter alone might (or might not) account for six times the amount of energy that would be added to the climate system by the doubling of atmospheric CO_2. Similar uncertainties attend other crucial variables like the impact of differing degrees of cloud cover and water vapor.

Given the room for error, it should come as no surprise that climate-prediction models have racked up an exceedingly poor track record over the years. According to those models, the average global temperature should have increased by at least 1 degree centigrade since the beginning of the 20th century, when industrial emissions of greenhouse gases first began to rise. But the best available measurements indicate that the average global temperature has increased by only 0.5 degrees in 100 years, and much of that increase occurred before 1940—too early in the century, in other words, to have been caused by a growth in CO_2 levels.

Contrary to the simulations, moreover, the marginal uptick in surface temperatures in the years since 1970 has not been accompanied by warming of the lower atmosphere (as we know from satellite data). A pair of recent papers in the journal *Science* attempts to account for this discrepancy by locating the missing heat in the oceans, a "discovery" trumpeted by the media as yet another blow to those who remain skeptical of global warming. But this was not a discovery at all, and was not based on any finding that whatever warming may have occurred has been caused by human activity. Rather, it was merely the product of "improved" models, which have their own "improved" assumptions and their own set of poorly understood parameters.

In the face of such scientific shell-games, and in the face of the huge costs the United States has been asked to incur to combat a problem that may or may not exist, President Bush was certainly right to pull the plug on the Kyoto Protocol. But whether he will be able to stand firm against the torrent of criticism that has been unleashed against him remains an open question. According to the Natural Resources Defense Council (NRDC), the Bush administration's decision to abandon Kyoto "will have massively destructive consequences for the earth and its people." Although the IPCC has specifically rejected any direct linkage between today's local environmental perturbations and global warming, the Sierra Club is instructing its members that the apocalypse is upon us now, in the form of "heat waves, droughts, coastal flooding, and malaria outbreaks."

Partisan politics

There are more narrowly partisan interests at play as well. "Democrats See Gold in Environment," ran the headline of a recent *New York Times* story describing how Bush's environmental decisions have galvanized activists in the Democratic party. Indeed, reports the *Times,* some party officials are positively "gleeful" at the political opportunities now opening up. One such official is evidently Senator Lieberman. Assuming Al Gore's mantle as the party's leading spokesman on matters environmental, Lieberman has called the decision to abandon Kyoto "flabbergasting," and is now invoking the specter of "sea levels [that] could swell up to 35 feet, potentially submerging millions of homes and coastal property."

That this is the same Joseph Lieberman who in 1997 joined 94 other Senators in voting to denounce the Kyoto Protocol suggests that when it comes to global warming we are indeed facing a rising tide—of hysteria, mixed with sheer political cynicism. As against these twin forces, it may seem hopelessly naive to suggest that we would do better to focus on phasing out those greenhouse gases that can be eliminated at relatively low cost, like sulfur hexafluoride and perfluorocarbons, while adopting a wait-and-see attitude toward CO_2, secure in the knowledge that advances in technology and in the accuracy of prediction will allow us to address climate change more effectively and more cheaply in the future. Naive it may be, but at present there is no basis in scientific evidence for more drastic action. All that is required is a politician tough enough and brave enough to say so.

5

Catastrophic Global Warming Is a Myth

Gary Benoit

Gary Benoit is editor of the New American, *a biweekly conservative journal.*

Thousands of respected scientists reject the theory that global warming is human-induced and will lead to disastrous climate change. Much of the mainstream media, however, ignore differences of opinion about climate among scientists and fail to report evidence proving that no catastrophic global warming is occurring. The global warming debate also tends to be monopolized by environmentalists with political agendas who want to convince the public that devastating climate change is imminent. In actuality, dire predictions about global warming are largely based on inaccurate computer models that do not take satellite data or other potentially significant factors into account.

Media reports to the contrary, President George W. Bush is concerned about the issue of global warming. Not as concerned as Bill Clinton or Al Gore. But concerned enough to deliver a speech on the subject.

"The issue of climate change respects no border," Bush warned on June 11th, 2001. "Its effects cannot be reined in by an army nor advanced by any ideology. Climate change, with its potential to impact every corner of the world, is an issue that must be addressed by the world."

"The Kyoto Protocol was fatally flawed in fundamental ways," Bush said. "But the process used to bring nations together to discuss our joint response to climate change is an important one. That is why I am today committing the United States of America to work within the United Nations framework and elsewhere to develop with our friends and allies and nations throughout the world an effective and science-based response to the issue of global warming."

The United Nations Kyoto (global warming) treaty, which Bush has rejected, at least for now, would require the industrialized nations to radically reduce their emissions of carbon dioxide (CO_2) and other "greenhouse" gases. This would mean a corresponding reduction in the use of fossil fuels. The United States, in order to implement Kyoto, would have to cut fossil fuel emissions by an estimated 30 percent. Such a cutback would be calamitous. "[I]f you like the California power crisis, you'll love Kyoto," Dr. S. Fred Singer, professor emeritus of environmental sciences at the University of Virginia, noted in an op-ed in the February 2nd, 2001, *Washington Times*.

Although Bush has backed away from Kyoto, he has not backed away from the issue of global warming or the supposed need to do something about it. ". . . America's unwillingness to embrace a flawed treaty should not be read by our friends and allies as any abdication of responsibility," he said in his June 11th speech. "To the contrary, my administration is committed to a leadership role on the issue of climate change." Like his predecessor, Bush is committed to reducing "greenhouse" gases—which is exactly what Kyoto was supposed to have accomplished. As Bush acknowledged: "Our approach must be consistent with the long-term goal of stabilizing greenhouse gas concentrations in the atmosphere."

The guardians of public opinion have greased the skids for Bush's adoption of that approach by portraying anything less than Kyoto as falling outside the mainstream. Yet there is no catastrophic global warming, and there is no need to reduce CO_2 emissions.

Media spin

There is, however, plenty of hot air. For instance, when the National Academy of Sciences (NAS) released a global warming report on June 6th, 2001, which was prepared at the request of the White House, the *New York Times* put the following spin on the story:

> A panel of top American scientists declared today that global warming was a real problem and was getting worse, a conclusion that may help the Bush administration alter its stand on the issue. . . .
>
> In a much-anticipated report from the National Academy of Sciences, 11 leading atmospheric scientists, including previous skeptics about global warming, reaffirmed the mainstream view that the earth's atmosphere was getting warmer and that human activity was largely responsible.

The *Times* pointed out that the authors of the report included "Dr. Richard S. Lindzen, a meteorologist at the Massachusetts Institute of Technology, who for years has expressed skepticism about some of the more dire predictions of other climate scientists about the significance of human-caused warming." But Dr. Lindzen has not reversed himself on the global warming issue, and he takes issue with the media's depiction of the report he coauthored.

In an op-ed appearing in the June 11th, 2001, *Wall Street Journal*, Dr. Lindzen noted how the NAS report "was depicted in the press as an implicit endorsement of the Kyoto Protocol. CNN's Michelle Mitchell was typical of the coverage when she declared that the report represented 'a unanimous decision that global warming is real, is getting worse, and is due to man. There is no wiggle room.'"

Dr. Lindzen continued: "As one of 11 scientists who prepared the report, I can state that this is simply untrue. For starters, the NAS never asks that all participants agree to all elements of a report, but rather that the report represent the span of views. This the full report did, making clear that there is no consensus, unanimous or otherwise, about long-term climate trends and what causes them." In fact, Lindzen pointed out, "the full text noted that 20 years was too short a period for estimating long-term trends, but the summary [which the media focused on] forgot to mention this."

The UN's Intergovernmental Panel on Climate Change (IPCC) is also cited by the media as representing a supposedly pro-Kyoto scientific consensus on the global warming issue. However, the IPCC's Summary for Policymakers, which is typically what the media uses as the basis for their reports, "represents a consensus of government representatives (many of whom are also their nations' Kyoto representatives), rather than of scientists," noted Dr. Lindzen.

But even full IPCC reports, which were written by scientists, were not immune from politics. In his testimony before the Senate Environment and Public Works Committee on May 2nd, Dr. Lindzen shed some light on the IPCC process: "The preparation of the report, itself, was subject to pressure. There were usually several people working on every few pages. Naturally there were disagreements, but these were usually hammered out in a civilized manner. However, throughout the drafting sessions, IPCC 'coordinators' would go around insisting that criticism of [computer-driven climate] models be toned down, and that 'motherhood' statements be inserted to the effect that models might still be correct despite the cited faults. Refusals were occasionally met with ad hominem attacks." Dr. Lindzen knows what he is talking about from direct experience, since he has been a participant in IPCC proceedings. "I personally witnessed coauthors forced to assert their 'green' credentials in defense of their statements," he said.

There is no consensus, unanimous or otherwise, about long-term climate trends and what causes them.

According to Dr. Lindzen, the IPCC's approach to the global warming issue is hardly surprising. "The IPCC was created to support the negotiations concerning CO_2 emission reductions" he said in his Senate testimony. "Although the press frequently refers to the hundreds and even thousands of participants as the world's leading climate scientists, such a claim is misleading on several grounds. First, climate science, itself, has traditionally been a scientific backwater. There is little question that the best science students traditionally went into physics, math and, more recently, computer science. Thus, speaking of 'thousands' of the world's leading climate scientists is not especially meaningful. Even within climate science, most of the top researchers (at least in the US) avoid the IPCC because it is extremely time consuming and non-productive. Somewhat ashamedly I must admit to being the only active participant in my department. None of this matters a great deal to the IPCC. As a UN activity, it is far more important to have participants from a hundred countries many of which have almost no efforts in climate research."

Scientific "skeptics"

Dr. Lindzen is often portrayed by the media as a global warming "skeptic," as if to suggest that his views fall outside the mainstream of scientific thought. In point of fact, thousands of scientists with impressive academic credentials do not buy the theory of catastrophic global warming. Many have made their views known publicly, only to be ignored by the major media. Dr. Lindzen, in fact, is just one of about 17,000 scientists who have signed an anti-Kyoto petition organized by Dr. Arthur Robinson, director of the Oregon Institute for Science and Medicine. Two-thirds of the scientists who signed this petition hold advanced degrees.

Dr. Frederick Seitz, past president of the National Academy of Sciences, wrote a cover letter asking scientists to sign the petition. "It is especially important for America to hear from its citizens who have the training necessary to evaluate the relevant data and offer sound advice," he explained. The petition itself states:

> We urge the United States government to reject the global warming agreement that was written in Kyoto, Japan, in December 1997, and any other similar proposals. The proposed limits on greenhouse gases would harm the environment, hinder the advance of science and technology, and damage the health and welfare of mankind.
>
> There is no convincing scientific evidence that human release of carbon dioxide, methane, or other greenhouse gases is causing or will, in the foreseeable future, cause catastrophic heating of the Earth's atmosphere and disruption of the Earth's climate. Moreover, there is substantial scientific evidence that increases in atmospheric carbon dioxide produce many beneficial effects upon the natural plant and animal environments of the Earth.

The widespread scientific support behind this statement dwarfs that of the IPCC's Summary for Policymakers. But that does not stop the media from claiming a nonexistent scientific consensus behind the theory of catastrophic global warming.

Facts, not fiction

Scientific conclusions should be based on observable facts, not political agendas. Yet politics is driving the global warming debate. "Science, in the public arena, is commonly used as a source of authority with which to bludgeon political opponents and propagandize uninformed citizens," Dr. Lindzen lamented in his *Wall Street Journal* article. "This is what has been done with both the reports of the IPCC and the NAS. It is a reprehensible practice that corrodes our ability to make rational decisions."

Yet rational decisions can be made. All that is necessary is to separate the politics from the science and examine the known facts:

• *Climate variability:* The climate is constantly changing, not just season to season but year to year, century to century, and millennium to millennium. In his *Journal* article, Dr. Lindzen pointed out that "two centuries ago, much of the Northern Hemisphere was emerging from a little ice age. A millennium ago, during the Middle Ages, the same region was in a warm period. Thirty years ago, we were concerned with global cooling." During the global cooling scare of the 1970s, some observers even worried that the planet was on the verge of a new ice age.

• *The actual temperature record:* The global mean temperature is approximately 0.5 degrees Celsius higher than it was a century ago. Based on surface readings, the temperature rose prior to 1940, perhaps in response to the end of the little ice age, which lasted until the 19th century. From about 1940 until about 1975, the temperature dropped, sparking the above-mentioned global cooling scare. More recently the temperature has been rising again, sparking concerns about global warming.

Thousands of scientists with impressive academic credentials do not buy the theory of catastrophic global warming.

The accuracy of the surface temperature record must be kept in mind when evaluating trends measured in fractions of a degree. One significant problem is the extent to which the data may be skewed as a result of urbanization. Atmospheric physicist Dr. S. Fred Singer wrote in a letter that appeared in the May 2001 issue of *Science:* "The post-1940 global warming claimed by the IPCC comes mainly from distant surface stations and from tropical sea surface readings, with both data sets poorly controlled (in both quality and location)." On the other hand, "surface data from well-controlled U.S. stations (after removing the urban 'heat-island' effects) show the warmest years as being around 1940." In his testimony to the Senate Commerce Committee on July 18th, 2000, Singer bluntly stated: "The post-1980 global warming trend from surface thermometers is not credible."

Dr. Singer, who established the U.S. Weather Satellite Service and served as its first director, is just one of many scientists who believe that temperature data collected by weather satellites provides a far better measuring stick than the surface readings. After all, the satellite data is truly global, and it is not skewed by the urban heat effect. The satellite data from January 1979 (when this data first became available) through May 2001 shows a warming trend of 0.038 degrees Celsius per decade—or less than four-tenths of one degree per century. This minuscule rate of increase, which could change, is far less than the dramatic increases in temperature the forecasters of doom have been warning against.

• *Man's effect on the climate:* In the interest of scrupulous accuracy, Dr. Lindzen acknowledged in his May 2nd Senate testimony that "man, like the butterfly, has some impact on climate." Obviously this was true when the Vikings were able to cultivate Greenland, Iceland, and Newfoundland. But it is true even today. In the April 3rd issue of the *Wall Street Jour-*

nal, George Melloan noted that, according to "serious scientists," "the greenhouse gases are a fundamental part of the biosphere, necessary to all life, and . . . industrial activity generates less than 5% of them, if that."

• *Carbon dioxide's effect on climate:* According to the global warming theory, the increase of carbon dioxide in the atmosphere, which has been established, is causing the global temperature to rise. Most of the increase in the surface temperature during the past century occurred before most of the increase in atmospheric CO_2. The temperature in 1940, recall, was not much different than it is now. Yet, as astrophysicist Sallie Baliunas pointed out in a letter published in the August 5, 1999 *Wall Street Journal,* "more than 80% of the manmade carbon dioxide has entered the air since the '40s."

One reason why the global warming theory may be flawed is that the amount of atmospheric CO_2 is not the only variable determining the earth's temperature. It is not even the main "greenhouse" gas. In a chapter appearing in the compendium *Earth Report 2000,* Dr. Roy Spencer, senior scientist for climate studies at NASA's Marshall Space Flight Center, noted: "It is estimated that water vapor accounts for about 95 percent of the earth's natural greenhouse effect, whereas carbon dioxide contributes most of the remaining 5 percent. Global warming projections assume that water vapor will increase along with any warming resulting from the increases in carbon dioxide concentrations."

The projected "positive feedback" to the initial CO_2-induced warming may not occur to the extent that global warming theorists are predicting, however. As Dr. Spencer points out, "there remain substantial uncertainties in our understanding of how the climate system will respond to increasing concentrations of carbon dioxide and other greenhouse gases." Moreover, the natural greenhouse effect that heats the earth is moderated by natural cooling processes. "In other words," concluded Dr. Spencer, "the natural greenhouse effect cannot be considered in isolation as a process warming the earth, without at the same time accounting for cooling processes that actually keep the greenhouse effect from scorching us all."

• *The sun's effect on climate*: One factor global warming theorists ignore is the effect that the sun's changing activity may have on the global temperature. A brighter sun may cause the global temperature to rise, and vice versa. Dr. Baliunas, in the *Wall Street Journal* letter referenced above, explained how the sun's activity can be measured by the length of the sunspot cycle (the shorter the cycle, the more active the sun). Dr. Baliunas' letter included a chart showing a close correlation between changes in the length of the sunspot cycle and Northern Hemisphere land temperature for 1750–1978.

Climate models

The known facts do not point to catastrophic global warming. That prediction is not based on the known temperature record but on complicated computer models that have been grossly inaccurate in the past. Those models do a very poor job of properly applying all the myriad factors that shape the world's climate, in large part because much of the mechanisms of climate remain largely unknown.

Dr. Frederick Seitz warned against relying on computer models of the climate in the *Wall Street Journal* for April 19th, 2001: "According to climate change models, the earth's surface temperature should have increased substantially in the past few decades because of man-made carbon dioxide already added to the atmosphere. However, actual temperature measurements show that these computer models have exaggerated the amount of warming by at least a factor of two." In light of this failure, Dr. Seitz reasoned: "Since the computer estimates of global warming for the past few decades have been cut back by a factor of two or more, to bring them in line with the measured temperature increases, the same correction should be applied to temperature predictions for the coming century. This would reduce the projected warming in 2100 to well within the range of natural variability of climate—the normal fluctuations that occur in nature without any human influence."

Dangerous solution

To head off the theoretical global warming threat, America and other developed nations are supposed to subject themselves to a global warming treaty that would result in an energy crisis so severe as to make California's energy shortfall appear mild by comparison. Full implementation of Kyoto would not save the earth from catastrophic global warming since no such threat exists. It would, however, reduce our standard of living and consolidate more power into the hands of those who intend to control and allocate the earth's supposedly limited resources.

It is not too surprising that the Clinton-Gore White House supported Kyoto, considering that administration's overt radicalism. Nor is it surprising that Clinton never submitted the Kyoto treaty to the Senate for ratification. He knew that the treaty would be dead on arrival, since that body had earlier voted 95-0 not to ratify any global warming treaty that did not include commitments on the part of developing nations such as India and China. What is surprising is that George W. Bush is now being cast as an anti-environment, anti-Mother Earth ignoramus for having criticized Kyoto *in its present form* when he should have stated that no global warming threat exists.

6

Global Warming Is Eroding Glacial Ice

Andrew C. Revkin

Andrew C. Revkin is a staff writer for the New York Times.

The recent melting of mountain glaciers indicates that contemporary global warming is largely the result of human activity. Although natural climatic changes are partly responsible for the rise in average temperatures over the past century, the current rate of glacial erosion suggests that higher levels of greenhouse gases—resulting from pollution—are contributing significantly to global warming. Glacial melting could lead to damaging flash floods in some locales. Other regions that depend on melting snow for hydroelectric power could run low on water after glaciers have disappeared, requiring more communities to use pollution-creating oil or coal for energy—which in turn would produce more greenhouse gases and more global warming.

The icecap atop Mount Kilimanjaro, which for thousands of years has floated like a cool beacon over the shimmering plain of Tanzania, is retreating at such a pace that it will disappear in less than 15 years, according to new studies.

The vanishing of the seemingly perpetual snows of Kilimanjaro that inspired Ernest Hemingway, echoed by similar trends on ice-capped peaks from Peru to Tibet, is one of the clearest signs that a global warming trend in the last 50 years may have exceeded typical climate shifts and is at least partly caused by gases released by human activities, a variety of scientists say.

Measurements taken over a year-long period on Kilimanjaro show that its glaciers are not only retreating but also rapidly thinning, with one spot having lost a yard of thickness between February 2000 and February 2001, said Dr. Lonnie G. Thompson, a senior research scientist at the Byrd Polar Research Center of Ohio State University.

Altogether, he said, the mountain has lost 82 percent of the icecap it had when it was first carefully surveyed, in 1912.

Given that the retreat started a century ago, Dr. Thompson said, it is likely that some natural changes were affecting the glacier before it felt any effect from the large, recent rise in carbon dioxide and other heat-trapping greenhouse gases from smokestacks and tailpipes. And, he noted, glaciers have grown and retreated in pulses for tens of thousands of years.

But the pace of change measured now goes beyond anything in recent centuries.

"There may be a natural part of it, but there's something else being superimposed on top of it," Dr. Thompson said. "And it matches so many other lines of evidence of warming. Whether you're talking about borehole temperatures, shrinking Arctic sea ice, or glaciers, they're telling the same story."

Glacial erosion

Dr. Thompson presented the fresh data in February 2001 at the annual meeting of the American Association for the Advancement of Science in San Francisco.

Other recent reports of changes under way in the natural world, like gaps in sea ice at the North Pole or shifts in animal populations, can still be ascribed to other factors, many scientists say, but many add that having such a rapid erosion of glaciers in so many places is harder to explain except by global warming.

The vanishing of the seemingly perpetual snows of Kilimanjaro . . . is one of the clearest signs that a global warming trend in the last 50 years may have exceeded typical climate shifts.

The retreat of mountain glaciers has been seen from Montana to Mount Everest to the Swiss Alps. In the Alps, scientists have estimated that by 2025 glaciers will have lost 90 percent of the volume of ice that was there a century ago. (Only Scandinavia seems to be bucking the trend, apparently because shifting storm tracks in Europe are dumping more snow there.)

But the melting is generally quickest in and near the tropics, Dr. Thompson said, with some ancient glaciers in the Andes—and the ice on Kilimanjaro—melting fastest of all.

Separate studies of air temperature in the tropics, made using high-flying balloons, have shown a steady rise of about 15 feet a year in the altitude at which air routinely stays below the freezing point. Dr. Thompson said that other changes could also be contributing to the glacial shrinkage, but the rising warm zone is probably the biggest influence.

Trying to stay ahead of the widespread melting, Dr. Thompson and a team of scientists have been hurriedly traveling around the tropics to extract cores of ice from a variety of glaciers containing a record of thousands of years of climate shifts. The data may help predict future trends.

The four-inch-thick ice cylinders are being stored in a deep-frozen archive at Ohio State, he said, so that as new technologies are developed for reading chemical clues in bubbles and water in ancient ice, there will still be something to examine.

The sad fact, he said, is that in a matter of years, anyone wanting to study the glaciers of Africa or Peru will probably have to travel to Columbus, Ohio, to do so.

Dr. Richard B. Alley, a professor of geosciences at Pennsylvania State University, said the melting trend and the link—at least partly—to human influence is "depressing," not only because of the loss of data but also because of the remarkable changes under way to such familiar landscapes.

"What is a snowcap worth to us?" he said. "I don't know about you, but I like the snows of Kilimanjaro."

Threatened water supplies

The accelerating loss of mountain glaciers is also described in a scientific report on the impact of global warming, which [was] released in February 2001 in Geneva by the Intergovernmental Panel on Climate Change, an influential network of scientists advising world governments under the auspices of the United Nations. The melting is likely to threaten water supplies in places like Peru and Nepal, the report says, and could also lead to devastating flash floods.

Kilimanjaro, the highest point in Africa, may provide the most vivid image of the change in glaciers, but, Dr. Thompson said, the rate of retreat is far faster along the spine of the Andes, and the consequences more significant. For 25 years, he has been tracking a particular Peruvian glacier, Qori Kalis, where the pace of shrinkage has accelerated enormously just since 1998.

From 1998 to 2000, the glacier pulled back 508 feet a year, he said. "That's 33 times faster than the rate in the first measurement period," he said, referring to a study from 1963 to 1978.

In the short run, this means the hydroelectric dams and reservoirs downstream will be flush with water, he said, but in the long run the source will run dry.

"The whole country right now, for its hydropower, is cashing in on a bank account that was built up over thousands of years but isn't being replenished," he said.

Once that is gone, he added, chances are that the communities will have to turn to oil or coal for power, adding even more greenhouse gases to the air.

In the Alps, scientists have estimated that by 2025 glaciers will have lost 90 percent of the volume of ice that was there a century ago.

The changes in the character of Kilimanjaro are registering beyond the ranks of climate scientists. People in the tourism business around the mountain and surrounding national park are worried that visitors will no longer be drawn to the peak once it has lost its glimmering cap.

Dr. Douglas R. Hardy, a geologist at the University of Massachusetts, returned from Kilimanjaro in February 2001 with the first yearlong record of weather data collected by a probe placed near the summit.

Just before he left, he had a long conversation with the chief ranger of Kilimanjaro National Park, who expressed deep concern about the trend. "That mountain is the most mystical, magical draw to people's imagination," Dr. Hardy said. "Once the ice disappears, it's going to be a very different place."

And the melting continues. When Dr. Hardy climbed the mountain to retrieve the data, he discovered that the weather instruments, erected on a tall pole, had fallen over because the ice around the base was gone.

7

Global Warming Threatens Arctic Life

Bruce E. Johansen

Bruce E. Johansen is the Robert T. Reilly Professor of Communications and Native American Studies at the University of Nebraska in Omaha. He is also the author of The Global Warming Desk Reference.

Global warming is having an adverse effect on the environment in the Arctic. Rising temperatures endanger populations of animals that live in an ice-based ecosystem. Polar bears, walruses, and seals, for example, are now often underweight or malnourished because higher temperatures limit their ice-based food sources and disrupt their feeding and weaning cycles. Furthermore, the native peoples of Alaska and Canada, many of whom depend on hunting and fishing for food and income, face increasing economic difficulties as Arctic game becomes more scarce.

It's another warm day in Iqaluit, capital of the new semi-sovereign Inuit nation of Nunavit in the Canadian Arctic. The bizarre weather is the talk of the town. The urgency of global warming is on everyone's lips.

While George W. Bush stalls with talk about needing "sound science," the temperature hit 82 degrees Fahrenheit on July 28, 2001, in this Baffin Island community that nudges the Arctic Circle. That's thirty-five degrees above the July average of 47, making it comparable to a 115- to 120-degree day in New York City or Chicago.

It is the warmest summer anyone in the area can remember. Swallows, sandflies, and robins are making their debuts, and pine pollen is affecting people as never before. Travelers joke about forgetting their shorts, sunscreen, and mosquito repellant—all now necessary equipment for a globally warmed arctic summer.

In Iqaluit (pronounced "Eehalooeet"), a warm, desiccating westerly wind raises whitecaps on nearby Frobisher Bay and rustles carpets of purple saxifrage flowers as people emerge from their overheated houses (which have been built to absorb every scrap of passive solar energy) with ice cubes wrapped in hand towels. The wind raises eddies of dust on Iqaluit's gravel roads as residents swat at the slow, corpulent mosquitoes.

An unmistakable reality

Welcome to the thawing ice-world of the third millennium. Around the Arctic, in Inuit villages connected by the oral history of traveling hunters as well as by e-mail now, weather watchers are reporting striking evidence that global warming is an unmistakable reality. Sachs Harbour, on Banks Island, above the Arctic Circle, is sinking into the permafrost. Shishmaref, an Inuit village on the far-western lip of Alaska sixty miles north of Nome, is being washed into the newly liquid (and often stormy) Arctic Ocean as its permafrost base dissolves.

In the Arctic, a world based on ice and snow is melting away.

"We have never seen anything like this. It's scary, very scary," says Ben Kovic, Nunavut's chief wildlife manager. "It's not every summer that we run around in our T-shirts for weeks at a time."

At 11:30 A.M. on a Saturday, Kovic is sitting in his backyard, repairing his fishing boat, wearing a T-shirt and blue jeans in the warm wind, with many hours of Baffin's eighteen-hour July daylight remaining. On a nearby beach, Inuit children are building sand castles with plastic shovels and buckets, occasionally dipping their toes in the still-frigid sea water.

"The glaciers are turning brown," he says, speculating that melting ice may be exposing debris and that air pollution may be a factor. Some ringed seals have been caught with little or no hair, he reports, though be doesn't have an explanation for this. "That is a big question that some-one has to answer," he says. Rivers have dried up that used to be spawn-ing grounds for Arctic char.

Other changes are more menacing. During Iqaluit's weeks of record heat in July, two tourists were hospitalized after they were mauled by a polar bear in a park south of town. On July 20, a similar confrontation oc-curred in northern Labrador as a polar bear tried to claw its way into a tent occupied by a group of Dutch tourists, according to the Toronto *Globe and Mail*. That time, the tourists escaped injury but the bear was shot to death.

In the Arctic, a world based on ice and snow is melting away.

The bears are "often becoming shore dwellers rather than ice dwellers," says Kovic. The harbor ice at Iqaluit did not form in the year 2000 until late December, five or six weeks later than usual. The ice also breaks up earlier in the spring, sometimes in May in places that once were icebound into early July. Polar bears usually obtain their food (seals, for example) from the ice. Without it, they can become hungry, miserable creatures, especially in unaccustomed warmth.

"The bears are looking for a cooler place," says Kovic.

On Hudson Bay in Manitoba, polar bears waking from their winter's slumber have found the ice melted earlier than usual. Instead of making their way onto the ice in search of seals, the bears walk along the coast until they get to towns like Churchill, where they block motor traffic and pillage the dump. Churchill now has a holding tank for wayward polar bears that is larger than its human jail.

Canadian Wildlife Service scientists reported in 1998 that polar bears around Hudson Bay were 90 to 220 pounds lighter than thirty years ago, apparently because earlier ice-melting has given them less time to feed on seal pups. When sea ice fails to reach a particular area, the entire ecological cycle is disrupted. When the ice melts prematurely, the polar bears can no longer use it to hunt for ring seals, many of which also have died, having had no ice to haul out on.

The offshore, ice-based ecosystem is sustained by upwelling nutrients, which feed the plankton, shrimp, and other small organisms, which feed the fish, which feed the seals, which feed the bears. Many Native people, who fish and hunt for their sustenance, are also deprived of a way of life. When the ice is not present, the entire cycle collapses.

Ice in many areas now melts earlier, sometimes as early as March, when the seals are having their pups. Because the ice breaks up too early, the pups often have not been fully weaned. Many of them starve or grow up in a weakened state. Warmer average temperatures may mean that the Arctic Ocean will become ice-free much of the year, imperiling populations of walrus and seal that feed on creatures living on the ice.

Changes in Inuit life

The Arctic's rapid thaw has made hunting, never a safe or easy way of life, even more difficult and dangerous. In the winter of 2000, Simon Nattaq, an Inuit hunter, fell through unusually thin ice and became mired in icy water long enough to lose both his legs to hypothermia, one of several injuries and deaths reported around the Arctic recently due to thinning ice.

Pitseolak Alainga, another Iqaluit-based hunter, said that climate change compels caution. One must never hunt alone, he said (Nattaq had been hunting by himself).

Alainga knows the value of safety on the water. His father and five other men died during late October 1994, after an unexpected storm swamped their hunting boat. The younger Alainga and one other companion barely escaped death in the same storm.

Within two or three generations, many Inuit have become urbanized as the tendrils of industrial life extend to the Arctic. Where visitors once arrived by dog sled or sailing ship, they now stream into Iqaluit's busy airport on Boeing 727s in which half the passenger cabin has been sequestered for freight. With no land-surface connections to the outside, freight as large as automobiles is sometimes shipped to Iqaluit by air.

The population of Iqaluit has jumped from about 3,500 to 6,500 in less than three years. Substantial suburban-style houses with mortgages worth hundreds of thousands of dollars have sprung up around town, rising on stakes sunk into the permafrost and granite hillsides. In other areas, ranks of walk-up apartments march along the high ridges above Frobisher Bay. Every ounce of building material has been imported from thousands of miles away.

People in Iqaluit subscribe to the same cable television services available in "the South." Bart Simpson and Tom Brokaw are well-known personages in Iqaluit, where some homes have sprouted satellite dishes. Iqaluit also now hosts a large supermarket of a size that matches stores in larger urban areas, except that the prices are three to four times higher than in Ottawa or Omaha. If one can afford the bill, mango-grapefruit juice and ready-cooked buffalo wings (as well as many other items of standard "southern" fare) are readily available.

Warmer average temperatures may mean that the Arctic Ocean will become ice-free much of the year, imperiling populations of walrus and seal that feed on creatures living on the ice.

Climate change has been rapid, and easily detectable within a single human lifetime. "When I was a child," said Sheila Watt-Cloutier, Canadian president of the Inuit Circumpolar Conference, "we never swam in the river [Kuujjuaq] where I was born [Nunavit, Northern Quebec], and now kids swim in there all of the time." Cloutier does not remember even having worn short pants as a child.

Gunter Weller, director of the Center for Global Change and Arctic System Research at the University of Alaska in Fairbanks, says mean temperatures in the state have increased by five degrees Fahrenheit in the summer and ten degrees in the winter over the last thirty years. Moreover, the Arctic ice field has shrunk by 40 percent to 50 percent over the last few decades and has lost 10 percent of its thickness, studies show.

"These are pretty large signals, and they've had an effect on the entire physical environment," Weller says.

Six hundred Native people in the village of Shishmaref are watching their village erode into the sea. The permafrost that had reinforced Shishmaref's waterfront is thawing. "We stand on the island's edge and see the remains of houses fallen into the sea," wrote Anton Antonowicz in the London *Daily Mirror* in the year 2000. "They are the homes of poor people. Half-torn rooms with few luxuries. A few photographs, some abandoned cooking pots. Some battered suitcases."

Percy Nayokpuk, a village elder, runs the local store, which now perches dangerously close to the edge of the advancing sea. "When I was a teenager, the beach stretched at least fifty yards further out," Percy told the *Daily Mirror*. As each year passes, the sea's approach seems faster. Five houses have washed into the sea; the U.S. Army has moved or jacked up others. The villagers have been told they will soon have to move. Year by year, the hunting season, which depends on the arrival of the ice, starts later and ends earlier.

"Instead of dog mushing, we have dog slushing." Clifford Weyiouanna told the paper.

In the Canadian Inuit town of Inuvik, ninety miles south of the Arctic Ocean near the mouth of the Mackenzie River, the temperature rose to 91 degrees on June 18, 1999.

"We were down to our T-shirts and hoping for a breeze," says Richard Binder, a local whaler and hunter. Along the Mackenzie River, according to Binder, "Hillsides have moved even though you've got trees on them. The thaw is going deeper because of the higher temperatures and longer periods of exposure."

In some places near Binder's village, the thawing earth has exposed ancestral graves, and the remains have had to be reburied.

Some Alaskan forests have been drowning and turning gray as thawing ground sinks under them. Trees and roadside utility poles, destabilized by thawing, lean at crazy angles. The warming has contributed a new phrase to the English language in Alaska: "the drunken forest."

Born in an igloo, Rosmary Kuptana grew up in Sachs Harbour. Now forty-seven, she has been an Inuit weather watcher for much of her life. Her job was to scan the morning clouds and test the wind's direction to help the hunters decide whether to go out, and what everyone should wear.

"We can't read the weather like we used to," says Kuptana. "The permafrost is melting at an alarming rate." Foundations of homes in Sachs Harbour are cracking and shifting. Kuptana says at least three experienced hunters recently fell to their deaths through unusually thin ice.

In the fall, storms have become more frequent and more violent, making boating difficult. Thunder and lightning have been seen for the first time, arriving with another type of weather that is new to the area: dousing summer rainstorms.

At Sachs Harbour, mosquitoes and beetles are now common sights; they were unknown a generation ago. Sea-ice is thinner and now drifts far away during the summer, taking with it the seals and polar bears the village's Inuit residents rely on for food.

"We have no other sources of food; the people in my community are completely dependent on hunting, trapping, and fishing," says Kuptana. "We don't know when to travel on the ice, and our food sources are getting farther and farther away. Our way of life is being permanently altered."

8

Global Warming Is Not a Threat to Polar Ice

Philip Stott

Philip Stott, emeritus professor of biogeography at London University, is coauthor of Political Ecology: Science, Myth, and Power.

Those who believe that human-caused global warming is a serious problem often point to changes in polar ice as proof of their theory. In 2002, for example, a 500-billion ton section of the Antarctic Larsen B ice shelf collapsed into the sea, prompting environmentalists to call for reductions in greenhouse gas emissions. However, shifting ice in the Antarctic provides no evidence of global warming. While some ice sheets have been retreating, others are growing thicker, and some Antarctic regions have actually become colder in recent years. The earth could, in fact, be on the verge of entering another ice age.

The dramatic demise of the Larsen B ice shelf in Antarctica in March 2002 has been embraced by environmentalists and commentators who warn of human-induced "global warming." After all, the ice shelf was 200 meters thick, with a surface area three times the size of Hong Kong. Around 500 billion tons of ice collapsed in less than a month. How could President George W. Bush ignore such evidence of our guilt with regard to climate change?

An ice shelf is a floating extension of the continental ice that covers the landmass of Antarctica. Larsen B was one of five shelves that have been monitored by scientists. The U.S.-based National Snow and Ice Data Center described its break-up as "the largest single event in a series of retreats by ice shelves in the peninsula over the last 30 years."

A simplistic myth

One worry can be dismissed immediately: Having been a shelf—a floating part of an ice sheet, rather than over land—it does not raise sea levels upon melting. Yet the collapse has proved to be a perfect natural disaster for the "Apocalypse Now" school of journalism. It is now perfectly clear that we are all doomed and that this is the wake-up call for urgent action on greenhouse gas emissions, automobiles, industry, and virtually everything else to do with economic growth.

Sorry, the North Pole isn't disappearing—and neither is the South Pole.

Unfortunately, the story isn't quite so straightforward. Antarctica illustrates the complexities behind understanding climate change, and it provides little support for a simplistic myth of human-induced "global warming." In fact this scare is reminiscent of a much-hyped *New York Times* story in 2001 that "leads" of open water in ice fields near the North Pole filled cruise passengers with a "sense of alarm" about impending climate disasters. But ice-breakers are always searching for "leads" to make their way through the ice, and after a long summer of 24-hour days it is not unusual to find them all over the place, especially after strong winds break up the winter ice. Sorry, the North Pole isn't disappearing—and neither is the South Pole.

Research on the West Antarctic Ice Sheet has shown precisely the opposite trend seen at Larsen B, namely that this ice sheet may be getting thicker, not thinner. Most scientists think that the sheet has probably been retreating, spasmodically, for around the last 10,000 years, but instead of the rate accelerating in recent years, it now appears to have halted its retreat. There is evidence that the ice sheet in the Ross Sea area is growing by as much as 26.8 gigatons per year, particularly on a part of the ice sheet known as Stream C.

This demonstrates the innate complexity of Antarctica as a continent. In reality, it has many "climates," and many geomorphological and glaciological regimes. It does not respond to change, whatever the direction, in a single, unitary fashion. Geomorphological and ecological trends are thus very difficult to interpret in a linear way.

A colder climate

One trend has been toward a colder climate. Over the last 50 years, the temperatures in the interior appear to have been falling. University of Illinois researchers have reported, in *Nature*, on temperature records covering a broad area of Antarctica. Their measurements show "a net cooling on the Antarctic continent between 1966 and 2000." Indeed, some regions, like the McMurdo Dry Valleys, the largest ice-free area, appear to have cooled between 1986 and 1999 by as much as two degrees centigrade per decade. As the researchers wryly comment, "Continental Antarctic cooling, especially the seasonality of cooling, poses challenges to models of climate and ecosystem change."

At the same time that parts of the continent are cooling, it's hardly surprising to see some ice melting. We are currently emerging—granted in a somewhat jerky fashion—out of the Little Ice Age that ended around 1880. It's to be expected that some parts of Antarctica like Larsen B are retreating. Yet we seem to be shocked at this perfectly natural event. When will we recognize the basic truth that change, both evolutionary and catastrophic, is the norm on our ever-restless planet?

Extreme environmentalists and sensationalist journalists pretend that every environmental event is of our own making. If only. We don't have that much control over Mother Nature. While we've been busy gabbing about global warming, the planet may be moving in the opposite direction.

Our current interglacial period is already 10,000 years old. No interglacial period during the last half-million years has persisted for more than 12,000 years. Most have had life spans of only 10,000 years or less. Statistically, therefore, we are due to slither into the next glacial period.

Despite a short-term rise in temperature of around 0.6 degrees centigrade over the last 150 years, the long-term temperature trend remains, overall, one of cooling. It may not be too long, therefore, before we see the ice spreading again. At worst, the emission of greenhouse gases is only likely to produce a super interglacial period; at worst, withdrawing gases might help to speed the descent into the next glacial period. And what would you prefer, a warmer or a colder world?

9

Global Warming Threatens World Health

Bruce Agnew

Bruce Agnew is a science writer based in Bethesda, Maryland.

Most scientists who have participated in recent climatological studies agree that global warming will affect the world's ecosystems and weather patterns. In addition, rising temperatures and increased precipitation are likely to have detrimental effects on human health. Extreme weather can cause injuries and deaths, interrupt food production, and contaminate water supplies. Changing climatic conditions might enable infectious tropical diseases such as malaria, encephalitis, and Ebola to spread into new geographical areas. Refugees fleeing from famine, rising sea levels, or weather-damaged regions could move into already crowded locales, straining public health and sanitation capabilities. The nations of the world must find ways to address these climate-related public health challenges.

M ost climatologists now believe that the Earth's atmosphere is warming, but no one knows how high, or how fast, temperatures may rise. And even though several national and international studies in 2001 predicted that tropical diseases such as malaria and dengue may extend their ranges as the world warms—and that disrupted storm and rainfall patterns may raise threats of everything from crop failures to cholera—no scientific consensus exists on precisely what ecological upsets will hit which countries, where, in the coming decades. Climate computer models cannot fine-tune their projections to regional levels that could tell local officials, for example, whether to prepare for droughts, or floods, or both.

But several major conclusions are clear. "What needs to be recognized is that there is very little doubt among leading scientists [who have taken part in recent studies] that climate change is a reality," says World Health Organization (WHO) environmental health expert Dr Carlos Corvalan. "We don't yet know how severe the impacts are going to be or how accurate the predictions of environmental change are, but the evidence is accumulating, and ecological and human health impacts are expected. We are also concerned that the health impacts of global warming will strike hardest at developing nations, particularly the poorest."

A nation's ability to adapt to climate change "depends on such factors as wealth, technology, education, information, skills, infrastructure, access to resources, and management capabilities," says the Third Assessment Report of the United Nations Intergovernmental Panel on Climate Change (IPCC), released early in 2001. "The developing countries, particularly the least developed countries, are generally poorest in this regard."

It is also clear that preparing for global warming is going to be an immensely complex task. Global warming "will require attention on many fronts," says Dr Jonathan Patz, director ʳ ʰ programme on health effects of global environmental ch ̄ ns Hopkins Bloomberg School of Public Health in ? SA.

In particular, global war. ew demands on public health officials and govern. tries, says Dr Bettina Menne, global change officer o₁ Centre for Environment and Health in Rome. Up ι most studies of the multiple, interlocking risk factors ' have been driven "not by the public health people ι .rʃ modellers, mathematicians and climatologists or econ. .e public health community must become more deeply involv .ιese assessments, she says.

At least, the modellers, mathematic. .₃ and climatologists have filled in the background. The IPCC's report projected that unless world governments take steps to stabilize emissions of carbon dioxide and other greenhouse gases, the global average surface temperature will rise by 1.4 degrees C to 5.8 degrees C (2.5 degrees F to 10.8 degrees F) between 1990 and 2100—a pace of warming that the report said is "very likely" unprecedented over the past 10,000 years.

The IPCC's Working Group I, which involved nearly 1000 scientists, predicted these changes: land areas will warm more rapidly than the oceans, particularly at high latitudes; precipitation will increase globally, with heavy precipitation over most land areas; in some areas precipitation will decline; and the sea level will rise by 9–88 centimetres between 1990 and 2100. "Extreme weather events"—such as heat-waves, heavy rains, floods, droughts, more ferocious hurricanes and typhoons, and drying out of soil at mid-latitudes—will likely increase, but current climate models cannot tell precisely where they will strike, the IPCC report said.

There is very little doubt among leading scientists . . . that climate change is a reality.

These projections are based on computer models that still have some gaps and uncertainties, but the scientific consensus supporting the forecast of a warmer world has become overwhelming. Even in the US, where global warming at times has been a political issue, the influential National Academy of Sciences signed on to the IPCC warming projections in June 2001. After a review requested by US President George Bush, a National Academy of Sciences committee reported: "The body of the [IPCC

Working Group I] report is scientifically credible and is not unlike what would be produced by a comparable group of only US scientists working with a similar set of emission scenarios, with perhaps some normal differences in scientific tone and emphasis."

Health effects of global warming

Assessing what global warming will mean for human health, however, is a hugely complex task, clouded by uncertainties. "One of the difficulties," says Patz, "is that we are talking about complex modes of exposure to the risk factors, and we're talking about long-term risk factors."

"If you raise the temperature a few degrees," Patz explains, "not only will that have an immediate physical effect on humans—especially on the elderly in urban areas—but raising the temperature changes atmospheric chemistry, which then can affect air pollution, especially tropospheric [low-level] ozone. Changes in temperature and precipitation can affect ecology and habitat for insect vectors of diseases. Warmer air holds more moisture, causing more extremes in the water cycle, giving you both droughts and flooding, affecting run-off and contamination, for example, from agriculture."

Heat-related deaths could rise in response to more frequent and more intense heat-waves, particularly in temperate-zone cities and among the elderly and urban poor.

Nevertheless, a series of international and national studies—and not a few individual scientists—have tried to puzzle out global warming's likely health effects. In addition to the IPCC effort, these include a WHO study in the year 2000 titled "Climate Change and Human Health: Impact and Adaptation", as well as government-sponsored national assessments in the UK, the US and several other countries. The US National Academy of Sciences, too, conducted a separate study on global warming and infectious diseases that was published in April 2001. Among the climate-triggered health threats that the studies spotlight are these:

• Vector-borne infectious diseases—such as malaria, dengue, schistosomiasis, leishmaniasis and encephalitis—may alter their geographical ranges and seasonality, spreading into new regions and declining in others. But some vector-borne disease experts say too many factors are involved in insect and disease organism life cycles to make projections based primarily on climatic changes.

• Heat-related deaths could rise in response to more frequent and more intense heat-waves, particularly in temperate-zone cities and among the elderly and urban poor who lack adequate air conditioning. But little research has been conducted on heat stress in developing countries, and scientists are only now beginning to examine heat morbidity—illness and disability short of death.

• Cold-related mortality might decline. In at least some temperate-zone countries, this reduction in cold-weather deaths might offset the increase in heat-stress mortality. But Johns Hopkins' Patz suspects that the lives saved wouldn't balance lives lost.

• Air pollution in urban areas would likely rise as air temperatures warm—particularly the concentration of ground-level ozone, which is damaging to respiratory health and is a main component of urban smog. At the same time, if current scientific understanding is correct, warming of the atmosphere at low levels would actually cool the stratosphere, accelerating the destruction of the stratospheric ozone that protects the planet from damaging ultraviolet radiation. Shifts in local weather also could alter regional pollution patterns and the spread of airborne allergens such as pollens and mould spores.

• Extreme weather events could "play a more significant role than even the warming itself in creating conditions conducive to outbreaks of disease," says Dr Paul Epstein, associate director of the Center for Health and the Public Environment at the Harvard Medical School in Boston, Massachusetts, USA. In addition to direct injury, and loss of life, violent weather can destroy shelter, contaminate water supplies, cripple food production, foster myriad infectious diseases, and tear apart existing health service infrastructures.

• Population displacement, forced by rising sea levels or extreme weather or agricultural collapse, would complicate the public health challenge. Large numbers of refugees moving into already populated areas, crowded together, hungry and perhaps starving, without shelter or adequate sanitation, is a formula for spreading infectious disease and promoting social conflict. "Personally I think that population displacement will be the iceberg under the tip of this problem," says Patz. "The displaced population issue could be the toughest and largest public health issue of climate change, yet it is without doubt the most difficult to put our arms around."

• Malnutrition risks, and the diseases that accompany malnutrition, would rise as agricultural practices adapt to new patterns of temperature, rainfall and soil-moisture conditions. Improved farm production in some regions, including northern Europe, might balance losses elsewhere. "But the risk of reduced food yields is greatest in developing countries—where 790 million people are estimated to be undernourished at present," the IPCC report says.

Violent weather can destroy shelter, contaminate water supplies, cripple food production, foster myriad infectious diseases, and tear apart existing health service infrastructures.

• Warming oceans could promote more frequent toxic algal blooms, increase the incidence of diarrhoeal diseases, and spread the risk of poisonings from fish and shellfish toxins that now are mostly limited to tropical waters.

• Emerging infectious diseases—not just known diseases such as Ebola haemorrhagic fever but also new diseases that science has not yet recognized—might be set free by ecosystem changes in response to shifting local weather conditions, providing new niches for non-native micro-organisms. Ecological systems that are upset might also spur the evolution of new strains of disease organisms, according to the US National Academy of Sciences study of linkages between climate, ecosystems and infectious disease in the United States.

"More people are expected to be harmed than benefited by climate change, even for global mean temperature increases of less than a few degrees," says the IPCC report. (And citizens of the poorer nations worst of all. Patz bristles at "the incredible inequity of this problem. The developed countries that are burning the most fossil fuel are the root of the problem, and yet it's the small island nations, the developing countries, that are really going to bear its brunt.")

Addressing the problem

So what's to be done?

At this early stage in science's understanding of global warming and its effects, no one seems to have a good answer to that question. Or at least, no one knows enough, yet, about the specific health problems that global warming may bring, to propose any detailed answers now. But there are a lot of wish-lists.

More research is first on every list: meteorological studies and development of better computer models to narrow down the specific, regional weather effects of climate change; improved surveillance of diseases like malaria and dengue, both to create a good database on their extent and to provide early warning of any spread of their ranges; new studies of the transmission dynamics of vector-, rodent- and waterborne diseases; and "integrative research" that takes into account the complex interactions within (and between) physical, ecological and societal systems that may make them vulnerable to climate change. The list of potential research subjects goes on and on.

"There's a lot of research that needs to be done, and some practical problems to study," says Harvard's Epstein. For example, "given what we already know about floods and mosquito-borne diseases, floods and cholera and waterborne diseases, there's a lot that we should be doing some real field work on."

It's also important now "to consider not just the potential impacts but to begin addressing adaptation measures," says WHO's Corvalan. "There's a realization that countries will need to take measures, as early as possible, to adapt to the potential changes, including changes to the health sector and delivery of health services. We need 'no-regrets' solutions, where benefits are achieved regardless of the magnitude of predicted impacts." This was the objective of a recent WHO workshop on small island countries organized in Samoa by Corvalan, Patz and Dr Hisashi Ogawa from WHO's Western Pacific regional office.

In addition, WHO's European regional office is conducting a three-year, 25-nation study of whether the preventive mechanisms are in place to cope with climate change. But Menne says the question of adaptation must be raised globally.

Adaptation will be costly. That's why developing nations are expected to have a harder time than the richer industrialized nations that can afford, and that already have, elaborate public health infrastructures. But there may be a silver lining.

"Most of the actions that are needed to adapt to the impacts of climate change—such as stepped-up vector-control efforts, improved water treatment systems and enhanced disaster-relief capability—would improve our health," says Dr Pim Martens, director of the Global Assessment Centre of Maastricht University's International Centre for Integrative Studies in the Netherlands, "even without global warming."

10

Global Warming Does Not Spread Tropical Diseases

H. Sterling Burnett and Merrill Matthews Jr.

H. Sterling Burnett is an environmental policy analyst. Merrill Matthews Jr. is vice president of domestic policy at the National Center for Policy Analysis in Dallas, Texas.

Global warming cannot be blamed for the spread of communicable tropical diseases. Prior to the nineteenth century, when the world was cooler, tropical diseases such as malaria and yellow fever abounded in the United States. In today's warmer world, however, malaria and yellow fever are rare in the United States—although similar illnesses are relatively common in Mexico. Overall, outbreaks of tropical disease have been on the decline globally. The epidemics that do occur in some regions are the result of poverty and poor medical care—not global warming.

Over the past few years the media have reported that one possible effect of global warming will be the expansion of tropical, communicable diseases borne by rodents or parasites into the United States. Fortunately, even if a warmer climate is in the offing, there is no reason for alarm, since the prime factor controlling communicable diseases is not global temperature, but relative wealth and the ecological and medical interventions people use to control diseases and their hosts.

Ground-level measurements of temperature show that the earth has warmed between 0.3 and 0.6 degrees Celsius the last 100 years.

Scientists at the World Health Organization (WHO) say that among the potentially deadly results of a warmer world are weather patterns that favor "opportunistic pests"—rodents, and parasitic insects—that often carry and transmit tropical diseases, including cholera, dengue fever, yellow fever and malaria. Scientists further hypothesize that as these pests enter new regions, the diseases they carry will spread.

According to a recent WHO report by Prof. Paul Epstein of the Harvard School of Public health, mosquitoes carrying malaria and dengue have been found at higher altitudes in Africa, Asia and Latin America due to warmer temperatures.

Scientists have noted that in 1995, the hottest year ever recorded, new outbreaks of dengue fever and the more deadly hemorraghic fever (DHF) occurred throughout Central America, where the number of dengue fever cases rose from 23,603 to 46,532 between September and November 1995.

In addition, in Mexico a 1995 outbreak of these diseases killed more than 4,000 people. And malaria expanded into new regions, becoming the ninth-leading cause of death in developing countries in 1990, claiming more than 856,000 lives.

Dr. Epstein contends that if the earth warms another 4 degrees Fahrenheit, the malaria-bearing mosquito's domain will expand from 42% of the globe to more than 60%.

While these and other tropical disease outbreaks are troubling, historical evidence and current medical data from the WHO indicate that they don't portend the spread of communicable diseases. A warm climate is a necessary condition for the mosquitoes that can carry malaria and dengue fever but is not a sufficient condition for the diseases to become epidemic.

Global warming is not promoting the spread of tropical diseases; bad policies are.

The climate in North America and most of Europe has long been suitable for the existence of several tropical diseases. From the 14th through the 19th Centuries, the world experienced a "little ice age," with temperatures in the Western hemisphere averaging 2 to 3 degrees Fahrenheit lower than current temperatures.

Despite the cooler temperatures, tropical diseases were fairly common in the United States. Indeed, malaria and yellow fever were endemic as far north as New York City, Philadelphia, Baltimore and Minnesota. In addition, between 1827 and 1946 there were eight pandemics of dengue fever, with more than 500,000 cases in 1922 alone. Malaria was widespread in the United States until quite recently, with more than 120,000 cases reported in 1934 and 63,000 cases in 1945.

Today, despite the warmer temperature, the U.S. rates of these diseases are quite low, even though the diseases are raging just beyond the U.S. border. For instance, when dengue fever struck Reynosa, Mexico, in 1995, there were 2,361 confirmed cases. By contrast, there were only 86 cases in the United States, all of which occurred in Texas, including 78 that were imported by immigrants arriving with the disease and only a couple of the nonimported cases were found in Hidalgo, just across the border from Reynosa.

Also in the United States, while ground-level temperature levels were increasing, malaria cases fell from 63,000 in 1945 to 2,000 in 1950 and 72 in 1960. Most new U.S. cases occur in immigrants, foreign travelers and U.S. citizens returning from travel in the tropics.

Further evidence that climate is not a primary cause of tropical diseases is the fact that the tropical city-state of Singapore reported no malaria deaths in 1994, while Malaysia, a nation that borders Singapore, suffered 36,853 cases of malaria, and Indonesia, the multi-island nation surrounding Singapore on three sides, reported 13,655 malaria cases.

Despite periodic regional outbreaks of communicable tropical diseases, tropical disease rates are decreasing on average globally, and at an even higher rate in developing countries. From 1983 to 1992, the most recent years for which firm data exist in Africa—the most heavily infested region of the world—reported cases of malaria fell from slightly under 3.2 million, after a brief intervening rise, to just over 420,000.

Bad policies

When outbreaks of communicable diseases occur, misguided policies are often the cause. For example, Peru had been cholera-free for many decades until 1991, due largely to chlorination of the drinking water supplies. But in 1991, based primarily on a study by the U.S. Environmental Protection Agency (EPA) showing that chlorine use posed a hypothetical increased risk of cancer, Peruvian officials ended their policy of water chlorination. As a result, more than 300,000 Peruvians contracted cholera the following year. The epidemic spread across South America, making more than 1 million people ill and taking more than 11,000 lives. In 1992, new research by the EPA determined that there was no link between cancer and chlorinated drinking water, but by then the damage had been done.

Other examples of misguided policies resulting in unnecessary illness and death abound. Through the use of DDT, malaria mortality in Ceylon fell from tens of thousands of cases to a few hundred each year. DDT was considered one of the safest pesticides in use. However, DDT was banned in the United States by the EPA for fear it was causing death and reproductive problems in bald eagles and other raptors—despite the fact that most scientists found no links between DDT use and thinner eggshells or bird deformities. Following the EPA's lead, U.S. aid agencies have refused funding to countries that continue to use DDT. Fearing a loss of the aid, Ceylon discontinued the use of DDT and malaria returned, taking thousands of lives each year.

Poverty promotes tropical disease

Almost every health expert recognizes that the prevalence of tropical diseases in the developing world stems from poverty and the conditions it entails, including lack of access to medical care and basic sanitation. In North America, cholera and yellow fever were virtually eradicated with the advent of filtered, chlorinated water and basic sanitation. Malaria was almost exterminated through a combination of judicious application of pesticides (primarily DDT), the draining of many of the nation's swamps, the use of screens on windows and doors, the rise of air-conditioner use and the widespread use of antimalarial drugs. Wherever this combination has been used, malaria rates have drastically declined.

Global warming is not promoting the spread of tropical diseases; bad policies are. Eliminate the policies that discourage economic growth, and regardless of the temperature, incomes will rise and tropical diseases will decline.

11

Global Warming Threatens the World Economy

Edward Goldsmith and Caspar Henderson

Edward Goldsmith is founder and publisher of The Ecologist, *Europe's leading environmental journal. He has also written several books critiquing industrialism and promoting environmental sustainability, including* The Case Against the Global Economy and for the Local. *Caspar Henderson is a journalist and consultant specializing in environment, security, and development.*

Climate change induced by global warming will devastate the world economy unless governments take measures now to reduce emissions of heat-trapping greenhouse gases. Rising sea levels could flood major coastal cities, displacing urban populations, exacerbating poverty and social disorder, and damaging the tourism industry. Extreme weather events, which are expected to become more frequent as global temperatures rise, could result in catastrophic property damage that would destroy the insurance industry. Since insurance companies are major investors in the world's financial markets, their collapse would adversely affect other sectors of the global economy, such as manufacturing. Industrialists who avoid facing the problem of global warming are contributing to their own demise.

Industrialists who continue to lobby governments to prevent them from taking the necessary action to combat climate change try to persuade themselves that inaction is in the best interests of their businesses and the economy itself. Given the enormous financial costs climate change will inflict, such an attitude is short-sighted in the extreme.

The first and most obvious way in which climate change will affect the economy is by the predicted sea-level rises. These . . . can increase from a mere 20 centimetres to several metres, depending on the effect of global warming on the Arctic and Antarctic ice-sheets. According to the Organisation for Economic Co-operation and Development (OECD), economic damages and losses arising from climatic destabilisation could cost the global economy up to $970 billion—on the basis of the present models which . . . tend to be optimistic. The opponents of appropriate preventive action must realise that a one-metre rise will be sufficient to flood most of New York City, including the entire subway system and all three major airports. New York, like many of the world's largest cities, is situated along the coast. The population densities of China's eleven coastal provinces average more than 600 people per square kilometre. Already nearly 40 per cent of the world's population lives within 100 kilometres of a coastline and more and more people are moving to coastal areas which are being increasingly degraded.

Let us not forget too, that the biggest industry of the world today is tourism. Most of it is in coastal areas and brings in billions of dollars in revenues every year. It would be foolish to suppose that tourism would not be affected by the consequent flooding of most of the beaches bordering tourist resorts, or by drastic heat-waves, water shortages and recurrent storms of greater and greater intensity, not to mention the effect on winter sports of retreating glaciers and ever thinner snow at ski resorts.

It would be equally foolish to suppose that the growing hordes of refugees will not affect the economy. It is only a question of time before state services are overwhelmed by a vast population of destitute people in the cities. The corresponding increase in crime and social disorder is very likely to interfere with commercial activities.

The effect on the insurance industry

The insurance industry is particularly vulnerable of course, and is becoming seriously concerned about what the future holds out for them. As Jeremy Leggett, who made a special study of this issue when he was Science Director of Greenpeace, notes, "Given only a slight increase in the scope for windstorms, drought-related wildfires, and floods, the $2 trillion insurance industry would be in danger of global collapse, with knock-on economic consequences which are completely ignored in most analyses of climate change." Property-catastrophe losses have already been enormous in recent years. 1992 was, at the time, the worst year ever, with global climatic natural catastrophe losses of over $22 billion, up 87 per cent on 1991, even allowing for inflation. 1993 was also a very bad year for disasters, especially flooding. In 1995, weather extremes caused $100 billion worth of damage, the highest figure ever, in 1996 the figure stood at $60 billion, and in 1998, costs to insurance companies rose to $90 billion, and it can only get worse.

A one-metre rise [in sea level] will be sufficient to flood most of New York City, including the entire subway system and all three major airports.

"Comparing the figures for the 1960s and the last ten years, we have established that the number of great natural catastrophes is three times larger," says Dr Gerhard Berz, Head of Geoscience Research at Munich Re, the world's largest reinsurer. "The cost to the world's economies after adjusting for inflation is nine times higher, and for the insurance industry three times as much."

According to research by Munich Re, there were more than 700 so-called 'large loss events' around the globe in 1998. These accounted for 85 per cent of economic losses and killed around 50,000 people. The most frequent natural catastrophes were windstorms, of which there were 240 significant ones, and floods, of which there were 170. In 1995, the previous most calamitous year, there were 100 fewer large loss events. In Britain the losses from flood damage for 1998 may top 1 billion pounds—"the worst floods anyone can remember, and happening twice within one year," says one observer.

If greenhouse gas emissions are allowed to continue to rise and global warming run its course, we will be facing by far and away the greatest catastrophe that our species has ever faced.

Munich Re clearly fingers global warming as the culprit for the extreme weather that has caused these mounting losses. Dr Gerhard Berz argues that a "further advance in man-made climate change will almost inevitably bring us increasingly extreme natural events and consequently increasingly large catastrophe losses."

Julian Salt, a disaster assessment expert at the Loss Prevention Council, says "the reinsurance pool contains between $200 billion and $300 billion. A couple of big storms in the wrong place—major cities on the US mainland, for example—could pretty much wipe that out." At the very least, he says, this would cause major dislocations to the world economy as insurers, facing heavy losses, pulled in their horns. Insurance companies are major investors in pension funds that contribute around a third of the capital in world financial markets. If they were to collapse then the effects on the economic system would be devastating.

Industrialists need to wake up

The insurance industry's dire prospects clearly augur ill for every sector of the economy, including manufacturing industry. Industrialists who still insist on opposing and preventing any action from being taken, on the grounds that it would cost too much, should enter the real world and wake up to the fact that the costs inflicted upon them through inaction will be enormous. If greenhouse gas emissions are allowed to continue to rise and global warming run its course, we will be facing by far and away the greatest catastrophe that our species has ever faced. Whatever may happen to the economy, what is absolutely certain is that we cannot live without a relatively stable climate and in particular one to which we and all the other forms of life with which we share this planet have been adapted by their co-evolution. To continue, therefore, to destabilise climate in order to satisfy what are referred to as economic requirements (but which in effect are those particular economic requirements needed to satisfy the immediate interests of the large transnational corporations that have come to dominate the economy), is at once an absurdity and a crime. Those who control these corporations, the governments, and the public at large, must recreate an economy that can function satisfactorily without disrupting our climate and indeed without continuing to pillage the natural world on whose integrity a stable climate ultimately depends.

12

Free Enterprise Must Be Confronted to Curb Global Warming

David Zink

David Zink is a contributor to Political Affairs, *a monthly journal published by the Communist Party.*

Scientific evidence suggests that human use of petroleum and coal creates heat-absorbing gases that contribute to global warming. Climatologists warn that over time, this warming will likely result in more frequent storms, rising sea levels, displacement of coastal populations and wildlife, and the spread of tropical diseases. However, corporations that profit from the sale of fossil fuels are attempting to censure these warnings by subsidizing a small group of politically influential scientists who deny the seriousness of global warming. The free enterprise system, with its emphasis on corporate profits, neglects the true needs of most of the world's people and of the global environment. The working class must unite to find a democratic way to allocate resources and preserve the environment.

Have you ever experienced a "Greenhouse Effect?" You have if you've ever opened the door of a car on a sunny day and the air inside feels hot enough to roast a turkey. Just as car windows allow sunlight in and trap the heat inside your car, certain gases allow sunlight in, then, instead of allowing the heat to reflect back into space, trap the heat within the Earth's atmosphere. For this reason, carbon dioxide (CO_2), methane, certain halogenated hydrocarbons, and nitrous oxide are termed "greenhouse gases."

This "greenhouse effect" is natural. Without it, the Earth would be a frozen rock spinning through space. Too much, and the Earth would be a barren rock spinning through space.

Seven of the hottest years on record occurred in the 1990s. Globally, the warmest year on record was 1998. Sea level rise of almost a foot in the last century has caused saltwater intrusion into many coastal area drinking water wells and destroyed beaches and wetlands around the world.

Polar and sub-Arctic regions have been experiencing warming well above the average for the past few decades. Thousands of square miles of Antarctic ice shelves collapsed and melted in the late 1990s. The area covered by sea ice decreased by about 6 percent from 1978–1995. Based on analysis of ice-core drilling samples from polar glaciers, scientists are saying that large bodies of open water right on the North Pole are appearing for the first time in 50,000 years. From the Himalayas to Mt. Kilimanjaro, and around the world, alpine glaciers are also shrinking.

Global climate change

To most people who have survived an icy northern winter, reports of a warming climate may come as good news. What can be bad about longer summers, and shorter, less severe winters? If that was all there was to it, this would be good news. But along with the warmer temperatures, climatologists are warning us to expect some unpleasant things like frequent and intense storms, a gradual rise in sea level, flooding at the coastlines, displacement of wildlife, hurricanes, hotter, drier summers, wetter winters, and the spread of tropical diseases. Some areas may actually get colder. For these reasons, the problem is better termed "Global Climate Change."

Studies have documented other dangers of climate change. Persistent neurological toxins such as DDT, banned in the US and Canada decades ago, are still used in many tropical countries. These chemicals evaporate into the atmosphere, return to earth in snowflakes that fall in higher altitudes, and are stored in glacial ice. Glacier melt increases the rate at which these compounds are released into the environment and concentrated in the food chain. These toxins are now turning up in high levels in the breast milk of Eskimo mothers.

Corporate-dominated globalization is the chief stumbling block to the resolution of environmental problems.

Harvard scientists have linked recent US outbreaks and the extension in the range of many insect- and water-borne diseases such as dengue ("breakbone") fever, malaria, Hantavirus, and schistosomiasis to the warming climate. No suitable vaccines exist for many of the diseases likely to spread as a result of the anticipated climate change.

By pumping greenhouse gases into the atmosphere and destroying the Earth's ability to cleanse itself and maintain climatic stability, capitalism is delivering a devastating one-two punch to our environment. Suburban sprawl is displacing natural vegetation and cropland at an alarming rate. Unlike forest and grassland, parking lots and strip malls don't absorb CO_2 or produce oxygen.

At the 1997 negotiations leading to the Kyoto Treaty, the UN set up an Intergovernmental Panel on Climate Change (IPCC) to investigate whether global warming is real or not and what dangers it might pose.

The IPCC commissioned 2,500 of the world's top climate scientists, including eight Nobel laureates to participate.

What conclusions did the IPCC reach?

1. There is evidence of a 25 percent increase in atmospheric CO_2 concentration since pre-industrial times.

2. Much evidence supports the view that global warming is caused by such human activities as fossil fuel combustion and clear-cutting and burning of forests.

3. Although there is uncertainty and disagreement on any future timetable of events, the long-term trend for an increase in both global temperature and sea surface temperature is unequivocal. The warming is projected to be greater than any seen in the last 10,000 years.

4. If current rates of fossil fuel burning continue, sea levels will rise between 15 to 95 centimeters.

Since the IPCC report is the best science we have available, prudence dictates that we use it as a guideline for the reduction of greenhouse gas emissions. Continued warming may mean a sea-level rise that would cause the flooding of Louisiana, Florida, Bangladesh, coastal cities like New York City, London, and Jakarta, small island nations, and other low-lying areas around the world. These areas are densely populated and climate changes could cause catastrophic population displacement.

The "debate"

What is the current status of the Kyoto Treaty? Island nations throughout the world are strong supporters. The European Union accepted the Kyoto Protocol and is trying to achieve lowered emissions. The treaty hasn't been submitted to the US Senate yet because of strong opposition. Key members have indicated that the treaty will be ratified over their dead bodies, because they say it would harm the economy. A Republican-controlled Senate would either vote it down or not let it out of committee. It will take 65 countries to ratify the treaty, but it won't be worth much if the US doesn't approve it.

The current debate is not over whether the climate is warming. Most scientists now agree on that. The question is now how quickly and how much the Earth is warming. The issue is complex because climate is affected by a mixture of natural events, such as volcanic activity, solar energy fluctuations and changes in oceanic water circulation. Dr. Michael E. Schlesinger, a climatologist at the University of Illinois at Urbana-Champaign, says it is important for policy makers and the public not to assume that temperature trends will follow a smooth rise. The array of factors is so complex that temporary fluctuations along an upward trend could create confusion and hamper work that could help solve the problem.

The Kyoto Treaty does have some serious shortcomings that illustrate how corporate-dominated globalization is the chief stumbling block to the resolution of environmental problems. Since most industrialized nations have fairly stringent controls on greenhouse gas emissions, the protocol will encourage corporate polluters to move to third-world locations where the treaty stipulates fewer or no controls. This will promote the migration of jobs to low-wage areas around the world. It is becoming increasingly evident that solving global environmental problems will require a better global economic system than capitalism.

Rather than admitting that our increasing use of petroleum products and coal is altering the climate, companies making profit from fossil fuels are spending millions of dollars in an effort to discredit the IPCC and global warming. Claiming that global warming is nothing but an "alarmist hoax," they have set out to buy the kind of "science" they want, and politicians are paying attention to these corporate-funded "climate experts."

The "Climate sceptics," a handful of scientists, directly subsidized by the fossil fuel lobby, are promoting blatant misinformation on climate science. Corporate-funded front groups base their arguments on a combination of deliberate misrepresentation of IPCC reports, contextual inaccuracy and unsubstantiated conclusions. They emphasize the disagreements at the Kyoto Conference, but don't talk about the overwhelming agreement on the basics. They say that even if global warming is real, so what? It is a good deal. Benefits, they say, include longer growing seasons, more precipitation and less stress on wildlife.

The "Climate sceptics," a handful of scientists, directly subsidized by the fossil fuel lobby, are promoting blatant misinformation on climate science.

This picture of a greener, more productive world is attractive, but would increased CO_2 concentrations really bring this about? While elevated CO_2 has been shown to increase plant growth and water-use efficiency under controlled conditions, doubt exists over how much of this benefit will be realized in the real world. We are living in a time of unprecedented clearing of forests and destruction of coral reefs. Burning fossil fuels is also the major cause of acid rain, urban smog, soot particles (which cause respiratory illnesses) and other environmental problems. The US, according to the Sierra Club, is the world's worst global warming polluter, accounting for 23 percent of emissions worldwide, while comprising only 4 percent of the world's population.

The real problem

Human-induced climate change is basically a pollution problem. Over its lifetime the average car today will spew 50 tons of CO_2 into the air. No solution to this problem will be found until we reduce automotive emissions and improve fuel efficiency, and/or increase the use of alternative technologies. The most important single step we can take to curb global warming is to make our cars go further on a gallon of gas. Raising the fuel economy standards in the US to 45 mpg for cars and 34 for light trucks would keep millions of tons of CO_2 out of the air. Cars, trucks, lawnmowers and power plants could be made more efficient by simply using better technology that already exists. Vehicles like the Toyota Prius and the Honda Civic VX have shown today's cars can get over 50 miles per gallon.

Instead of yielding to the power of advertising, "People should buy a vehicle appropriate to their needs," says Dan Becker, director of the Sierra

Club's global warming program. They shouldn't buy one that climbs mountains and pulls yachts [and gets only 24 miles per gallon or less] if they're just going to use it to drive to work and pick up the kids from school."

Gasoline engines are no longer even necessary. We have the technology to move to solar energy for electricity and hydrogen-burning engines that produce only water vapor as exhaust. Cleaner, renewable sources of energy are available, affordable and waiting to be developed. Adopting this technology in the US, according to the American Council for an Energy-Efficient Economy, would be great for the economy, create 770,000 jobs, save $530 per household per year and significantly reduce the threat of global warming.

Eighty percent of greenhouse gases are produced by only 122 corporations.

So what's stopping us? These common-sense measures run smack into powerful corporate interests. A recent report from the Natural Resources Defense Council and the Union of Concerned Scientists points out that 80 percent of greenhouse gases are produced by only 122 corporations. Exploitation of fossil fuels is easier and far more lucrative for the oil companies than the development of solar or wind power. These corporations are jeopardizing the integrity of the entire global ecosystem, and holding the world's people and governments hostage by a combination of bribery and brute force.

At present, governments are using taxpayers' money to carry out the corporate agenda. Development of solar, wind and other clean energy is neglected. Instead, our tax dollars are subsidizing the fossil fuel industry, the chief source of greenhouse gas emissions. At the same time, the federal government is selling logging rights to our national forests to timber companies at the rate of pennies per tree. What are the results? Cheap timber for corporations at the expense of CO_2-absorbing trees, the loss of wildlife habitat, recreation areas, and air and water quality.

A widespread realization is growing that we need to start treating our life-support system with more respect. Because of the high priority it places on short-term corporate profits, capitalism tends to exacerbate the tendency to push the need to preserve a healthy environment off the stage.

Under the rules of the capitalist system, corporations are compelled to maximize gains and minimize costs—or lose to the competition. They do this by privatizing gains and externalizing costs to the commons. The environment has served as a free sewer to dump corporate wastes. Costs of cleanup are a public cost, financed by our tax dollars. Meanwhile, the wealthy and their corporations have successfully lobbied over the years to lower their share of the tax burden, shifting it onto the backs of the working class.

Capitalism is acting as a cork on the teapot of human creativity. If something is profitable to corporations, it happens, even if it is detrimental to the vast majority of people, our communities, and our global life-support system. But if something is not seen as profitable, even if it would be beneficial to the majority, then the corporations just aren't interested. The command decisions regarding which technologies are utilized are made by CEOs in the narrow interests of major corporate stockholders, not by those who do the work and who must live with the consequences of those decisions.

The solution

Environmental education is important in instilling an environmental ethic in our children. But corporate-financed critics are saying that it is an effort to brainwash our kids and fosters a "gloom and doom" perspective. They make their most scathing attacks on the science of climate change, ozone depletion, deforestation, species extinction and acid rain. They stress that environmental educators ought to be teaching that the free enterprise system provides the best environmental protection. According to the National Audubon Society, an estimated two million teachers have received this kind of material from various companies and trade associations and dollar-strapped schools are increasingly accepting corporate-developed curricula. This tends to have a chilling effect on the willingness of teachers, school administrators and parents to push for sound environmental education curricula.

It is the so-called free-enterprise system that got us into this mess to begin with! Let us be clear on this: enterprise can only be truly free when we, the working-class majority, join together and democratically decide what is produced, how it is to be produced and how the rewards are to be allocated. Only then can we disempower the parasites who systematically steal the wealth labor creates and wreck the environment we all depend upon.

We need to move beyond paycheck issues and transform our unions into class and environmentally-conscious organizations that are also concerned with control over such issues as technological change to enhance, not damage, environmental health. We need to build on the experience of Seattle in 1999, when workers and environmentalists joined forces as allies in struggle against arrogant transnational corporations. Sooner or later—and the sooner the better—we need to take the struggle into our workplaces, to operate the offices and factories for ourselves. For it is only in that way that we'll be able to start building a cooperative economic democracy that will fundamentally change this world for the better.

13

Global Warming Will Be Beneficial

Thomas Gale Moore

Thomas Gale Moore is a senior fellow at Stanford University's Hoover Institution.

Global warming will benefit humanity. Fewer people will die from cold-related complications, heating costs will decline, and increased plant growth will help to feed a growing world population. Rather than struggling to reduce greenhouse gas emissions—which would be financially burdensome—industrialized nations should promote economic growth and prosperity in poor countries so that they will be better equipped to handle potentially disruptive changes in climate.

In the recent debates over global warming in Washington and other world capitals, a critical point has been lost: Global warming, if it were to occur, would probably benefit most Americans. If humankind had to choose between a warmer or a cooler climate, we would certainly choose the former. Humans, nearly all other animals, and most plants would be better off with higher temperatures. The climate models suggest, and so far the record confirms, that, under global warming, nighttime winter temperatures would rise the most and daytime summer temperatures, the least. Most Americans prefer a warmer climate to a colder one—and that preference is justified. More people die of the cold than of the heat. More die in the winter than the summer. Statistical evidence suggests that the climate predicted for the end of the twenty-first century might reduce U.S. deaths by about forty thousand annually.

In addition, less snow and ice would reduce transportation delays and accidents. A warmer winter would cut heating costs, more than offsetting any increase in air-conditioning expenses in the summer. Manufacturing, mining, and most services would be unaffected. Longer growing seasons, more rainfall, and higher concentrations of carbon dioxide would benefit plant growth. Already there is evidence that trees and other plants are growing more vigorously. Although some locales may become too dry, too wet, or too warm, on the whole humankind should benefit from an upward tick in the thermometer.

Economic effects

What about the economic effects? In the pessimistic view of the Inter-governmental Panel on Climate Change (IPCC), the costs of global warming might be as high as 1.5 percent of the U.S. gross domestic product by the end of the twenty-first century. The cost of reducing carbon dioxide emissions, however, would be much higher. William Cline of the Institute for International Economics has calculated that the cost of cutting emissions by one-third of current levels by the year 2040 would be 3.5 percent of worldwide gross domestic product. The IPCC also reviewed various estimates of losses from stabilizing emissions at 1990 levels, a more modest objective, and concluded that the cost to the U.S. economy would be at least 1.5 percent of gross domestic product by 2050, with the burden continuing to increase thereafter.

The forecast cost of warming is for the end of the twenty-first century, not the middle. Adjusting for the time difference, the cost to the United States from a warmer climate at midcentury, according to the IPCC, would be at most 0.75 percent of gross domestic product, meaning that the costs of holding carbon dioxide to 1990 levels would be twice the gain from preventing any climate change. But the benefit-cost calculus is even worse. And if Third World nations, such as China, India, and Brazil, are exempted from the requirements of the treaty, Americans would pay a huge price for virtually no benefit.

> *On the basis of the evidence, including historical records, global warming is likely to be good for most of humankind.*

And even if the developing countries agreed to return emissions to 1990 levels, greenhouse gas concentrations would not be stabilized. The buildup would only slow because for many decades more carbon dioxide would be added to the atmosphere than removed through natural processes; consequently temperatures would continue to go up. Instead of saving the full 0.75 percent of gross domestic product by keeping emissions at 1990 levels, we would be saving much less.

Whatever dangers global warming may pose, they will be most pronounced in the developing world. It is much easier for rich countries to adapt to any long-term shift in weather than it is for poor countries, which tend to be much more dependent on agriculture. Poor countries lack the resources to aid their flora and fauna in adapting, and many of their farmers earn too little to survive a shift to new conditions. But the best insurance for these poor countries is an increase in their wealth, which would diminish their dependence on agriculture and make it easier for them to adjust to changes in weather, including increases in precipitation and possible flooding or higher sea levels. Subjecting Americans to high taxes and onerous regulations will help neither poor countries—we could buy less from them—nor us.

The optimal way to deal with potential climate change is not to embark on a futile attempt to prevent it but to promote growth and prosperity so that people will have the resources to deal with the normal set of natural disasters. On the basis of the evidence, including historical records, global warming is likely to be good for most of humankind. The additional carbon, rain, and warmth should promote the plant growth necessary to sustain an expanding world population. Global change is inevitable. Warmer is better. Richer is healthier.

Organizations to Contact

The editors have compiled the following list of organizations concerned with the issues debated in this book. The descriptions are derived from materials provided by the organizations. All have publications or information available for interested readers. The list was compiled on the date of publication of the present volume; the information provided here may change. Be aware that many organizations take several weeks or longer to respond to inquiries, so allow as much time as possible.

Center for Global Change Science (CGCS)
77 Massachusetts Ave., MIT 54-1312, Cambridge, MA 02139
(617) 253-4902 • fax: (617) 253-0354
e-mail: cgcs@mit.edu • website: http://web.mit.edu

CGCS at the Massachusetts Institute of Technology addresses long-standing scientific problems that impede accurate predictions for changes in the global environment. The long-term goal of CGCS is to accurately predict environmental changes by utilizing scientific theory and observations to understand the basic processes and mechanisms controlling the global environment. The center publishes and distributes a Report Series of papers intended to communicate new results and provide reviews and commentaries on the subject of global climate change.

Climate Solutions
610 Fourth Ave. E, Olympia, WA 98501
(360) 352-1763 • fax: (360) 943-4977
e-mail: info@climatesolutions.org • website: http://climatesolutions.org

Climate Solutions' mission is to stop global warming at the earliest possible point by helping the northwest region of the Unites States develop practical and profitable solutions. It focuses on job creation, economic development, and environmental protection. Climate Solutions publishes reports covering the impact of global warming, including *Global Warming Is Here: The Scientific Evidence*, available from its website.

Competitive Enterprise Institute (CEI)
1001 Connecticut Ave. NW, Suite 1250, Washington, DC 20036
(202) 331-1010 • fax: (202) 331-0640
e-mail: info@cei.org • website: www.cei.org

CEI is a nonprofit public policy organization dedicated to the principles of free enterprise and limited government. CEI encourages the use of private incentives and property rights to protect the environment. Instead of government regulation, it advocates removing governmental barriers and establishing private sector responsibility for the environment. CEI's publications include the monthly newsletter *CEI Update*, *On Point* policy briefs, and the books *The True State of the Planet* and *Earth Report 2000*.

The George C. Marshall Institute
1730 K St. NW, Suite 905, Washington, DC 20006
(202) 296-9655 • fax: (202) 296-9714
e-mail: info@marshall.org • website: www.marshall.org

The institute is a nonprofit research group that provides scientific and technical advice and promotes scientific literacy on matters that have an impact on public policy. It is dedicated to providing policy makers and the public with rigorous, clearly written, and unbiased technical analyses of public policies. The institute publishes several studies on global warming, including *A Scientific Discussion of Climate Change* and *Are Human Activities Causing Global Warming?*

Global Warming International Center (GWIC)
22W381 75th St., Naperville, IL 60565
(630) 910-1551 • fax: (630) 910-1561
website: www.globalwarming.net

GWIC is an international body that disseminates information on global warming science and policy to governmental and nongovernmental organizations and industries in more that 120 countries. The center sponsors unbiased research supporting the understanding of global warming and its mitigation. GWIC publishes the quarterly newsletter *World Resource Review.*

Goddard Institute for Space Studies (GISS)
2880 Broadway, New York, NY 10025
(212) 678-5641
e-mail: emichaud@giss.nasa.gov • website: www.giss.nasa.gov

GISS, a division of the National Aeronautics and Space Administration (NASA), is an interdisciplinary research initiative addressing natural and man-made changes in the environment that affect the habitability of the planet. A key objective of GISS research is the prediction of atmospheric and climate changes in the twenty-first century. The institute publishes numerous research papers on global warming that are available from its website.

The Heartland Institute
19 South LaSalle #903, Chicago, IL 60603
(312) 377-4000 • fax: (312) 377-5000
e-mail: think@heartland.org • website: www.heartland.org

The Heartland Institute is a nonprofit public policy research organization that provides research and commentary to elected officials, journalists, and its members. The institute conducts policy studies on global warming and other environmental issues. It publishes the monthly *Environment & Climate News* newspaper and the book *Eco-Sanity: A Common-Sense Guide to Environmentalism.*

The Heritage Foundation
214 Massachusetts Ave. NE, Washington, DC 20002
(202) 546-4400 • fax: (202) 546-8328
e-mail: info@heritage.org • website: www.heritage.org

The Heritage Foundation is a conservative think tank that supports free enterprise and limited government in environmental matters. Its publications, such as the quarterly *Policy Review*, the *Backgrounder*, and the *Heritage Lectures*, include studies on the uncertainty of global warming and the greenhouse effect.

The Intergovernmental Panel on Climate Change (IPCC)
C/O World Meteorological Organization, 7bis Ave. de la Paix, C.P. 2300,
CH-1211 Geneva 2, Switzerland
+41-22-730-8208 • fax: +41-22-730-8025
e-mail: ipcc_sec@gateway.wmo.ch • website: www.ipcc.ch

Recognizing the problem of potential global climate change, the World Meteorological Organization and the United Nations Environment Programme established the IPCC in 1988. The IPCC's role is to assess the scientific, social, and economic information relevant for the understanding of the risk of human-induced climate change. The IPCC has published its Third Assessment Report, *Climate Change 2001*, in addition to special reports on global warming.

Pew Center on Global Climate Change
2101 Wilson Blvd., Suite 550, Arlington, VA 22201
(703) 516-4146 • fax: (703) 841-1422
website: www.pewclimate.org

The Pew Center on Global Climate Change is a nonprofit, nonpartisan, and independent organization dedicated to educating the public and key policy makers about the causes and potential consequences of global climate change. By releasing reports on environmental impacts, policy issues, and economics, the center works to encourage the domestic and international communities to reduce emissions of greenhouse gases. Its reports include *The Science of Climate Change, Human Health and Climate Change,* and *Sea-Level Rise and Global Climate Change.*

Reason Foundation
3415 Sepulveda Blvd., Suite 400, Los Angeles, CA 90034
(310) 391-2245 • fax: (310) 391-4395
e-mail: keng@reason.org • website: www.reason.org

Reason Foundation is a national public policy research organization that supports the rule of law, private property, and limited government. It believes that choice and competition will achieve the best outcomes in social and economic interactions. The foundation specializes in a variety of policy areas, including the environment, education, and privatization. It publishes the monthly magazine *Reason* and the studies *Evaluating the Kyoto Approach to Climate Change* and *Global Warming: The Greenhouse, White House, and Poorhouse Effects.*

Sierra Club
85 Second St., Second Floor, San Francisco, CA 94105
(415) 977-5500 • fax: (415) 977-5799
e-mail: information@sierraclub.org • website: www.sierraclub.org

The Sierra Club is a grassroots organization that promotes the protection and conservation of natural resources. In addition to numerous books and fact sheets, the Sierra Club publishes the bimonthly magazine *Sierra*, the *Environmental Currents* newsletter, and special reports, including *Driving Up the Heat: SUVs and Global Warming.*

Union of Concerned Scientists (UCS)
2 Brattle Square, Cambridge, MA 02238
(617) 547-5552 • fax: (617) 864-9405
e-mail: ucs@ucsusa.org • website: www.ucsusa.org

UCS works to advance responsible public policy in areas where science and technology play a vital role. Its programs focus on safe and renewable energy technologies, transportation reform, arms control, and sustainable agriculture. UCS publications include the quarterly magazine *Nucleus,* the quarterly newsletter *earthwise,* and the global warming reports *Greenhouse Crisis: The American Response* and *A Small Price to Pay: U.S. Action to Curb Global Warming Is Feasible and Affordable.*

World Resources Institute (WRI)
10 G St. NE, Suite 800, Washington, DC 20002
(202) 729-7600 • fax: (202) 729-7610
e-mail: lauralee@wri.org • website: www.wri.org

WRI provides information, ideas, and solutions to global environmental problems. Its mission is to encourage society to live in ways that protect earth's environment for current and future generations. The institute's program attempts to meet global challenges by using knowledge to catalyze public and private action. WRI publishes the reports *Climate, Biodiversity, and Forests: Issues and Opportunities Emerging from the Kyoto Protocol* and *Climate Protection Policies: Can We Afford to Delay?*

Worldwatch Institute
1776 Massachusetts Ave. NW, Washington, DC 20036
(202) 452-1999 • fax: (202) 296-7365
e-mail: worldwatch@worldwatch.org • website: www.worldwatch.org

The Worldwatch Institute is dedicated to fostering the evolution of an environmentally sustainable society in which human needs are met in ways that do not threaten the health of the natural environment or the prospects of future generations. The institute conducts interdisciplinary and nonpartisan research on emerging global environmental issues such as climate change, the results of which are widely disseminated throughout the world. It publishes the annual *State of the World* anthology, the bimonthly magazine *World Watch,* and *Slowing Global Warming: a Worldwide Strategy* from the Worldwatch Paper Series.

Bibliography

Books

John J. Berger

Beating the Heat: Why and How We Must Combat Global Warming. Berkeley, CA: Berkeley Hills Books, 2000.

Donald A. Brown

American Heat. Lanham, MD: Rowman and Littlefield, 2002.

W. Bradnee Chambers

Inter-Linkages: The Kyoto Protocol and the International Trade and Investment Regimes. New York: University Press, 2001.

Gale E. Christianson

Greenhouse: The 200-Year Story of Global Warming. New York: Walker and Company, 1999.

Francis Drake

Global Warming: The Science of Climate Change. New York: Oxford University Press, 2000.

Dinyar Godrej

The No-Nonsense Guide to Climate Change. New York: Verso, 2001.

Kenneth Green

Global Warming: Understanding the Debate. Berkeley Heights, NJ: Enslow, 2002.

Bruce E. Johansen

The Global Warming Desk Reference. Westport, CT: Greenwood Publishing Group, 2001.

Jeremy Leggett

The Carbon War: Global Warming at the End of the Oil Era. London: Penguin, 2000.

M. Mihkel Mathieson and Zbigniew Jaworowski

Global Warming in a Politically Correct Climate: How Truth Became Controversial. Lincoln, NE: iUniverse, 2000.

Patrick J. Michaels and Robert C. Balling Jr.

The Satanic Gases: Clearing the Air About Global Warming. Washington, DC: Cato Institute, 2000.

Thomas Gale Moore

Climate of Fear: Why We Shouldn't Worry About Global Warming. Washington, DC: Cato Institute, 1998.

National Assessment Synthesis Team

Climate Change Impacts on the United States: The Potential Consequences of Climate Variability and Change. Washington, DC: U.S. Global Change Research Program, 2000.

S. Fred Singer

Global Climate Change: Human and Natural Influences. New York: Paragon House, 1989.

David G. Victor

The Collapse of Kyoto Protocol and the Struggle to Slow Global Warming. NJ: Princeton University Press, 2001.

Periodicals

Amicus Journal

"High Noon," Winter 2001.

Seth Borenstein

"Experts See Global Warming Disaster," *San Diego Union-Tribune*, February 19, 2001.

Kevin Burke	"Global Warming Bedtime Story," *In These Times*, January 8, 2001.
George W. Bush	"Global Climate Change," *Vital Speeches*, July 1, 2001.
William H. Calvin	"The Great Climate Flip-Flop," *Atlantic Monthly*, January 1998.
Eugene Linden Churchill	"The Big Meltdown," *Time*, September 4, 2000.
Jean-Michel Cousteau	"Before the Ice Melts," *Skin Diver*, May 2001.
Thomas J. Crowley	"Causes of Climate Change Over the Past 1000 Years," *Science*, July 14, 2000.
Gary Cuneen	"It's Time to Wake Up and Smell the Carbon," *U.S. Catholic*, November 2001.
Ross Gelbspan	"Reality Check," *E: The Environmental Magazine*, September/October 2000.
James K. Glassman	"Global Climate Scare: Fools Rush In," *Reason Online*, October 9, 2000.
Fred Guterl	"The Truth About Global Warming," *Newsweek International*, July 23, 2001.
Bob Herbert	"Rising Tides," *New York Times*, February 22, 2001.
Robert Lee Hotz	"Melting Releases Riddle on Global Warming," *Los Angeles Times*, April 1, 2001.
Richard Kerr	"Rising Global Temperatures, Rising Uncertainty," *Science*, April 13, 2001.
Paul Kingsnorth	"Human Health on the Line," *The Ecologist*, March/April 1999.
Joel Kovel	"Global Warming and Realo-Fundi Greens," *Z Magazine*, February 2001.
Gordon Laird	"Losing the Cool," *Mother Jones*, March/April 2002.
Michael D. Lemonick	"Life in the Greenhouse," *Time*, April 9, 2001.
Bill McKibben	"Too Hot to Handle," *New York Times*, January 5, 2001.
Robert Mendelsohn	"The Peculiar Economics of Global Warming," *Milken Institute Review*, Second Quarter 2000.
Patrick J. Michaels	"Global Warming Warnings: A Lot of Hot Air," *USA Today Magazine*, January 2001.
Nick Middleton	"The Heat Is On," *Geographical*, January 2000.
Lee Morrison	"If the World Really Is Getting Warmer, Humanity Should Learn to Adapt," *The Report Newsmagazine*, March 19, 2001.
National Catholic Reporter	"Ignoring Evidence of Global Warming a Risky Gamble," January 11, 2002.
Sid Perkins	"Big Bergs Ahoy!" *Science News*, May 12, 2001.

| Charles W. Petit | "Polar Meltdown," *U.S. News & World Report*, February 28, 2000. |

S. George Philander — "A Global Gamble," *Tikkun*, November 1999.

Kelly Reed — "To the Extreme," *Greenpeace*, Winter 2000.

Arthur B. Robinson and Noah Robinson — "Some Like It Hot," *American Spectator*, April 2000.

Stephen H. Schneider and Kristin Kuntz-Duriseti — "Facing Global Warming," *World & I*, June 2001.

Nancy Shute et al. — "The Weather Turns Wild," *U.S. News & World Report*, February 5, 2001.

S. Fred Singer — "Global-Warming Theory Steams Ahead Despite Conflicting Evidence," *Insight*, February 26, 2001.

Gar Smith — "W2K: The Extreme Weather Era," *Earth Island Journal*, Summer 2000.

Chuck Sudetic — "The Good News About Global Warming," *Rolling Stone*, October 12, 2000.

Ralph de Toledano — "The Enviros: Some Notes for the Record," *Insight*, December 24, 2001.

Robert W. Tracinski — "The Guilt of the Scientists," *Intellectual Activist*, July 2001.

Index